Everything You Ever Wanted to Know About DVD

Everything You Ever Wanted to Know About DVD

The Official DVD FAQ

Jim Taylor

McGraw-Hill
New York   Chicago   San Francisco   Lisbon
London   Madrid   Mexico City   Milan   New Delhi
San Juan   Seoul   Singapore   Sydney   Toronto

The **McGraw·Hill** *Companies*

Library of Congress Cataloging-in-Publication Data

Taylor, Jim (Jim H.)
    Everything you ever wanted to know about DVD : the official DVD FAQ /
    Jim Taylor.
      p. cm.
    ISBN 0-07-142038-X
     1. DVDs. I. Title.

TK7882.D93.T39 2003
621.388'332 — dc21
2003053997

1 2 3 4 5 6 7 8 9 0   DOC/DOC   0 9 8 7 6 5 4 3

ISBN 0-07-142038-X

*The sponsoring editor for this book was Steve Chapman and the production supervisor
was Pamela Pelton. It was set in Helvetica by MacAllister Publishing Services, LLC.*

Printed and bound by RR Donnelley.

 This book is printed on recycled, acid-free paper containing a minimum of 50
percent recycled de-inked fiber.

McGraw-Hill books are available at special quantity discounts to use as premiums and sales pro-
motions, or for use in corporate training programs. For more information, please write to the Director of Special Sales, McGraw-Hill Professional, Two Penn Plaza, New York, NY 10121-2298. Or
contact your local bookstore.

DEDICATION

This book is dedicated to the original members of the alt.video.dvd newsgroup and to readers of my DVD FAQ everywhere (except for those of you who sent e-mail asking me for a detailed description of what DVD is, who invented it, how it works, and how it will impact society, ASAP, so you can finish your school report due tomorrow).

Contents

Preface

In 1996, when those of us in the alt.video.dvd Internet newsgroup were collectively trying to figure out what DVD was going to be like and how it would affect the future, I volunteered to create a FAQ (frequently asked questions) document. Little did I know at the time that the book I was writing, *DVD Demystified*, would become "the bible" of the DVD industry and stretch into at least three editions. At first I asked hundreds of questions, which many very generous people answered. I collected and distilled the answers into the DVD FAQ. As time passed and my collection of DVD information increased, I discovered that I now knew the answers to almost all the questions, including the ones that arrived in my e-mail inbox with increasing frequency as more and more people encountered DVD. In the seven years since I started the DVD FAQ, I've worked hard to keep it up-to-date as the ultimate repository of information about DVD. I honestly don't know how many thousands of hours I've spent doing this. I've never tried to figure it out for fear of realizing that I really need to get a different hobby, such as fishing or collecting cereal boxes. However, my efforts do seem to be appreciated—the DVD FAQ is now accessed by hundreds of thousands of people, providing information at the rate of over 50 gigabytes each month from dvddemystified.com, not to mention dozens of copies at other web sites and translated versions in 11 languages.

It's an odd thing to turn a living electronic document into a printed book, but we haven't yet gotten to the point where anyone can browse web pages on an airplane or in a waiting room or at a video shoot in the middle of the desert. So if this book reaches more people and helps them better understand and use DVD, then it has been well worth my time and effort. I hope it answers every one of your questions.

Chapter 1

General DVD

What Is DVD?

DVD is the latest generation of optical disc storage technology. A DVD is essentially a bigger, faster *compact disc* (CD) that can hold cinema-like video, better-than-CD audio, still photos, and computer data. DVD aims to encompass home entertainment, computers, and business information with a single digital format. DVDs have replaced laserdiscs, are well on the way to replacing videotape and video game cartridges, and could eventually replace audio CDs and CD-ROMs. DVD has widespread support from all major electronics companies, computer hardware companies, and movie and music studios. With this unprecedented support, DVDs has become the most successful consumer electronics product of all time in less than three years after arriving on the market. In 2003, six years after their first appearance, over 250 million DVD playback devices were in operation worldwide, counting DVD players, DVD PCs, and DVD game consoles. This was more than half the number of VCRs, setting DVD up to become the new standard for video publishing.

It's important to understand the difference between the *physical formats,* such as DVD-ROM and DVD-R, and the *application formats,* such as DVD-Video and DVD-Audio. DVD-ROM is the base format that holds data. DVD-Video (often simply called DVD) defines how video programs such as movies are stored on disc and played in a DVD-Video player or a DVD computer. (See "Can I Play DVD Movies on My Computer?" in Chapter 4, "DVDs and Computers."). The situation is similar to the differences between CD-ROMs and audio CDs.

DVD-ROMs include recordable variations: *DVD-R,* DVD-RW, DVD-RAM, DVD+R, and DVD+RW. (See "What About Recordable DVDs: DVD-R, DVD-RAM, DVD-RW, DVD+RW, and DVD+R?" in Chapter 4.) The application formats include *DVD-Video, DVD-Video Recording* (DVD-VR), *DVD+RW Video Recording* (DVD+VR), *DVD-Audio Recording* (DVD-AR), *DVD Stream Recording, DVD-Audio* and *Super Audio CD* (SACD). Also, special application formats are used for game consoles such as Sony PlayStation 2 and Microsoft Xbox.

What Does DVD Stand For?

All of the following phrases have been proposed as the meaning behind DVD:

- Delayed, very delayed (referring to the many late releases of DVD formats)
- Diversified, very diversified (referring to the proliferation of recordable formats and other spin-offs)
- Digital venereal disease (referring to the piracy and copying of DVDs)
- Dead, very dead (from naysayers who predicted DVD would never take off)
- Digital video disc (the original meaning proposed by some of the creators)
- Digital versatile disc (a meaning later proposed by some creators)
- Nothing

And the official answer is? "Nothing." The original meaning was *digital video disc*. Some members of the DVD Forum have pointed out that DVD goes far beyond video and have offered the painfully contorted phrase *digital versatile disc* as a solution, but this has never been officially accepted by the DVD Forum. (See "Who Invented DVD, Who Owns It, and Whom Should Be Contacted for Specifications and Licensing?" in Chapter 6, "Miscellaneous.") The DVD Forum decreed in 1999 that DVD, as an international standard, is simply three letters. After all, how many people ask what VHS stands for? (Guess what? No one agrees on that one either.)

What Are the Features of DVD-Video?

The most important features of DVD are as follows:

- Over 2 hours of high-quality digital video (A double-sided, dual-layer disc can hold about 8 hours of high-quality video, or 30 hours of VHS-quality video.)
- Support for widescreen movies on standard or widescreen TVs (4:3 and 16:9 aspect ratios)
- Up to eight tracks of digital audio (for multiple languages, commentaries, and so on) that each have as many as eight channels

- Up to 32 subtitle/karaoke tracks
- Automatic, seamless video branching (for multiple story lines or ratings on one disc)
- Up to nine camera angles (different viewpoints can be selected during playback)
- Onscreen menus and simple interactive features (for games, quizzes, and so on)
- Multilingual identifying text for title name, album name, song name, cast, crew, and so on
- Instant rewind and fast forward (no "be kind, rewind" stickers on rental discs)
- Instant search to title, chapter, music track, and timecode
- Durable (no wear from playing, only from physical damage)
- Not susceptible to magnetic fields, resistant to heat
- Compact size (discs are easy to handle, store, and ship, and players can be portable; replication is cheaper than tapes or laserdiscs.)
- Noncomedogenic

NOTE: Most discs do not contain all features (multiple audio/ subtitle tracks, seamless branching, and parental control) because each feature must be specially authored. Some discs may not allow searching or skipping.

Most players support a standard set of features:

- Language choice (for the automatic selection of video scenes, audio tracks, subtitle tracks, and menus)*
- Special effects playback: freeze, step, slow, fast, and scan
- Parental lock (for denying playback of an entire disc or particular scenes of objectionable material)*
- Programmability (playing selected sections in a desired sequence)
- Random play and repeat play
- Digital audio output PCM stereo and Dolby Digital

---

*Must be supported by additional content on the disc

- Recognition and output of *Digital Theater Systems* (DTS) Digital Surround audio tracks

- Playback of audio CDs

Some players include additional features:

- Component video output (YUV or RGB) for a higher-quality picture

- Progressive-scan component output (YUV or RGB) for highest-quality analog picture

- Digital video output (SDI, 1394, or DVI] for perfect digital picture

- Six-channel analog output from internal audio decoder

- Playback of Video CDs or Super Video CDs, MP3 CDs, Picture CDs and Photo CDs, laserdiscs, and CDVs

- Reverse single-frame stepping

- Reverse play (normal speed)

- RF output (for TVs with no direct video input)

- Multilingual onscreen display

- Multiple-disc capacity

- Digital zoom (2x or 4x enlargement of a section of the picture. This is a player feature, not a DVD disc feature.)

What's the Quality of DVD-Video?

DVDs have the capability to produce near-studio-quality video and better-than-CD-quality audio. They are vastly superior to consumer videotape and generally better than laserdiscs (see "How Does DVD Compare to Laserdisc?" in Chapter 2, "DVD's Relationship to Other Products and Technologies"). However, quality depends on many production factors. As compression experience and technology improves, we see increasing quality, but as production costs decrease and DVD authoring software becomes widely available, we also see more shoddily produced discs. A few low-budget DVDs even use MPEG-1 encoding (which is no better than VHS) instead of higher-quality MPEG-2.

DVD-Video is usually encoded from digital studio master tapes to the MPEG-2 format. The encoding process uses *lossy* compression that removes redundant information (such as areas of the picture that don't change) and information that's not readily perceptible by the human eye. The resulting video, especially when it's complex or changes quickly, may sometimes contain visual flaws, depending on the processing quality and amount of compression. At average video data rates of 3.5 to 6 *megabits per second* (Mbps), *compression artifacts* may be occasionally noticeable. Higher data rates can result in higher quality, with almost no perceptible difference from the master at rates above 6 Mbps. As MPEG compression technology improves, better quality is being achieved at lower rates.

Video from DVD sometimes contains visible *artifacts* such as color banding, blurriness, blockiness, fuzzy dots, shimmering, missing detail, and even effects such as a face that "floats" behind the rest of the moving picture. It's important to understand that the term artifact refers to anything that is not supposed to be in the picture. Artifacts are sometimes caused by poor MPEG encoding, but they are more often caused by a poorly adjusted TV, bad cables, electrical interference, sloppy digital noise reduction, improper picture enhancement, poor film-to-video transfers, film grain, player faults, disc read errors, and so on. Most DVDs exhibit few visible MPEG compression artifacts on a properly configured system. If you think otherwise, you are misinterpreting what you see.

Some early DVD demos were not very good, but this simply indicates how bad DVD can be if not properly processed and correctly reproduced. In-store demos should be viewed with a grain of salt, because most salespeople are incapable of properly adjusting a TV set.

Most TVs have the sharpness set too high for the clarity of DVDs. This exaggerates high-frequency video and causes distortion, just as the treble control set too high on a stereo causes the audio to sound harsh. For best quality, the sharpness control should be set very low. Brightness should also not be set too high. Some DVD players output video with a black-level setup of 0 IRE (the Japanese standard) rather than 7.5 IRE (the U.S. standard). On TVs that are not properly adjusted, this can cause blotchiness in dark scenes. An option may be offered in the player menu to output standard black level. DVD-Video has exceptional color fidelity, so muddy or washed-out colors are almost always a problem in the display (or the original source), not in the player or disc.

DVD audio quality is superb. DVDs include the option of PCM digital audio with sampling sizes and rates higher than audio CDs. Alternatively, audio for most movies is stored as discrete, multichannel surround sound using Dolby Digital or DTS audio compression similar to the digital surround sound formats used in theaters. As with video, audio quality depends on how well the processing and encoding has been done. In spite of compression, Dolby Digital and DTS can be close to or better than CD quality.

What Are the Disadvantages of DVD?

The downsides to DVDs are as follows:

- The vagueness of the DVD specification and the inadequate testing of players and discs has resulted in incompatibilities. Some movie discs don't function fully (or don't play at all) on some players. (See "Why Doesn't Disc X Work in Player Y?" in this chapter.)

- DVD recorders are more expensive than VCRs. (See "Can DVDs Record from a VCR, TV, and So On?" and Chapter 4's "What About Recordable DVDs: DVD-R, DVD-RAM, DVD-RW, DVD+RW, and DVD+R?")

- DVD has built-in copy protection and regional lockouts. (See "What Are Regional Codes, Country Codes, or Zone Locks?" and "What Are Copy Protection Issues?" in this chapter)

- DVD uses digital compression. Poorly compressed audio or video may be blocky, fuzzy, harsh, or vague. (See "What's the Quality of DVD-Video?")

- The audio downmix process for stereo or Dolby Surround may reduce the dynamic range. (See "What Are the Audio Details?" in Chapter 3, "DVD Technical Details.")

- DVD doesn't fully support *high-definition TV* (HDTV). (See "Does DVD Support HDTV (DTV)? Will HDTV Make DVD Obsolete?" in Chapter 2.)

- Some DVD players and drives can't read CD-Rs. (See "IS CD-R Compatible with DVD?" in Chapter 2.)

- Some DVD players and drives can't read recordable DVDs. (See "Is It True There Are Compatibility Problems with Recordable DVD Formats?" in Chapter 4.)

- Most DVD players and drives can't read DVD-RAM discs. (See "DVD-RAM" in Chapter 4.)

- Very few players can play in reverse at normal speed.

- Variations and options such as DVD-Audio, DVD-VR, and DTS audio tracks are not supported by all players.

Which DVD Players and Drives Are Available?

Some manufacturers originally announced that DVD players would be available as early as the middle of 1996. These predictions were woefully optimistic. Delivery was initially held up for "political" reasons, which meant movie studios were demanding copy protection, and it was delayed again by a lack of titles.

The first players appeared in Japan in November of 1996, followed by U.S. players in March of 1997, with distribution limited to only seven major cities for the first six months. Players slowly trickled in to other regions around the world. Prices for the first players in 1997 were $1,000 and up. By the end of 2000, players were available for under $100 at discount retailers. In 2003, players became available for under $50. Six years after the initial launch, close to 1,000 models were available from over 100 consumer electronics manufacturers (see "Who Is Making or Supporting DVD Products?" in Chapter 6).

Fujitsu supposedly released the first DVD-ROM-equipped computer in November 1996 in Japan. In early 1997, Toshiba released a DVD-ROM-equipped computer and a DVD-ROM drive in Japan (moved back from December of 1996, which was moved back from November). DVD-ROM drives from Toshiba, Pioneer, Panasonic, Hitachi, and Sony also began appearing as manufacturer samples as early as January 1997, but none were available before May. The first PC upgrade kits (a combination of DVD-ROM drive and hardware decoder card) became available from Creative Labs, Hi-Val, and Diamond Multimedia in April and May of 1997.

Today, every major PC manufacturer has models that include DVD-ROM drives. The price difference from the same system with a CD-ROM drive ranges from $30 to $200 (laptops have more expensive drives). For more information about DVDs on computers, including writable DVD drives, see Chapter 4.

NOTE: If you buy a player or drive from outside your country (such as a Japanese player for use in the United States), you may not be able to play region-locked discs on it. (See "What Are Regional Codes, Country Codes, or Zone Locks?")

Pioneer released the first DVD-Audio players in Japan in late 1999, but they did not play copy-protected discs. Matsushita (under the Panasonic and Technics labels) first released full-fledged players in July 2000 at prices from $700 to $1,200. DVD-Audio players are now also made by Aiwa, JVC, Kenwood, Madrigal, Toshiba, Yamaha, and others. Sony released the first SACD players in May 1999 for $5,000. Pioneer's first DVD-Audio players, released in late 1999, also played SACD. SACD players are now also made by Accuphase, Aiwa, Denon, Kenwood, Marantz, Philips, Sharp, and others. (See "What About DVD-Audio or Music DVDs?" for more information on DVD-Audio and SACD.)

The following web sites provide more information on players and drives:

- CNET DVD players and DVD-ROM drives, http://computers.cnet.com
- The uk.media.dvd FAQ, www.dvd.reviewer.co.uk/umdvdfaq
- Aus.dvd (Australia/New Zealand/region 4 player info), www.ozemail.com.au/~brierley/dvd
- Computer Shopper DVD players and DVD-ROM drives, http://zdentshopper.cnet.com

Which Player Should I Buy?

The video and audio performance of all modern DVD players is excellent. Your personal preferences, budget, and home theater setup all play a large role in determining which player is best for you. Unless you have a high-end home theater, a player that costs under $250 should be completely adequate.

Make a list of things that are important to you (such as the ability to play CD-Rs, the ability to play Video CDs, 96 kHz/24-bit audio decoding, DTS Digital Out, and an internal 6-channel Dolby Digital decoder) to help you come up with a set of players. Then try out a few of the players in your price range, focusing on ease of use (remote control design, user interface, and front-panel controls). Because few variations exist in picture and sound quality within a given price range, convenience features play a big part. The remote control, which you'll use all the time, can drive you crazy if it doesn't suit your style.

Some players, especially cheaper models, don't properly play all discs. Before buying a player, you may want to test it with a few complex DVDs, such as *The Matrix*, *The Abyss*, *Independence Day*, and *DVD Demystified*. See "Why Doesn't Disc X Work in Player Y?" for more information.

In certain cases, you might want to buy a DVD PC instead of a standard DVD player, especially if you want progressive video. (See "What's a Progressive DVD Player?" and Chapter 4's "Can I Play DVD Movies on My Computer?")

Here are a few questions to ask yourself:

- Do I want selectable soundtracks and subtitles, multiangle viewing, aspect ratio control, and parental or multirating features? Do I want fast and slow playback, great digital video, multichannel digital audio, and compatibility with Dolby Pro Logic receivers? Do I also want onscreen menus, dual-layer playback, and the ability to play audio CDs? This is a trick question, because all DVD players have all these features.

- Do I want DTS audio? If so, look for a player with the DTS Digital Out logo. (See "Audio Details of DVD-Video," in Chapter 3.)
- Do I want to play Video CDs? If so, check the specs for Video CD compatibility. (See "Are Video CDs Compatible with DVD Players?" in Chapter 2.)
- Do I want to play recordable DVDs? If so, check the specs or compatibility reports for the ability to read -R, -RW, +R, and +RW formats. (See "Is It True There Are Compatibility Problems with Recordable DVD Formats?" in Chapter 4.)
- Do I need a headphone jack?
- Do I want player setup menus in languages other than English? If so, look for a multilanguage setup feature. (Note: All players support on-disc multilanguage menus.)
- Do I want to play homemade CD-R audio discs? If so, look for the dual laser feature. (See "Is CD-R Compatible with DVD?" in Chapter 2.)
- Do I want to replace my CD player? If so, you might want a changer that can hold three, five, or even hundreds of discs.
- Do I want to play discs from other countries? If so, beware of regions (see "What Are Regional Codes, Country Codes, or Zone Locks?") and TV formats. (See "Is DVD-Video a Worldwide Standard? Does It Work with NTSC, PAL, and SECAM?")
- Do I want to control all my entertainment devices with one remote control? If so, look for a player with a programmable universal remote, or make sure your existing universal remote is compatible with the DVD player.
- Do I want to zoom in to check details of the picture or get rid of the black letterbox bars? If so, look for players with picture zoom.
- Do I have a DTV or progressive-scan display? If so, get a progressive-scan player. (See "What's a Progressive DVD Player?")
- Do I want to play high-definition compatible digital (HDCD)s? If so, check for the HDCD logo. (See "Is HDCD Compatible with DVDs?" in Chapter 2.)
- Does my receiver have only optical or only coax digital audio inputs? If so, make sure the player has outputs to match. (See "How Do I Hook Up a DVD Player?" in Chapter 3.)

- Do I care about black-level adjustment?

- Do I value special deals? If so, look for free DVD coupons and rentals available with many players.

For more information, read hardware reviews at Web sites such as *DVD-File* or in magazines such as *Widescreen Review*. You may also want to read about user experiences at *Audio Review* and in online forums at *Home Theater Forum* and *DVDFile*. More advice can be found at *DVDBuyingGuide* and at *eCoustics.com*, which also has a list of links to reviews on other sites.

See Chapter 3's "What Are the Outputs of a DVD Player?" and "How Do I Hook up a DVD Player?" and for specific information on which audio/video connections are needed for your existing setup.

Which DVD Titles Are Available?

In the video distribution industry, a *title* refers to a movie or other production release, such as *Snow White, Star Wars,* or a boxed edition of a TV series like *Babylon 5 First Season*. Titles are collectively referred to as *software,* not to be confused with computer software.

DVDs started off slowly. In 1996, rosy predictions of hundreds of movie titles being sold for Christmas failed to materialize. Only a handful of DVD titles, mostly music videos, were available in Japan for the November 1996 launch of DVDs. The first feature films on DVD appeared in Japan in December, including *The Assassin, Blade Runner, Eraser,* and *The Fugitive* from Warner Home Video. By April, over 150 titles were available in Japan. The first titles released in the United States on March 19, 1997, by Lumivision, authored by AIX Entertainment, were IMAX adaptations: *Africa: The Serengeti, Antarctica: An Adventure of a Different Nature, Tropical Rainforest,* and *Animation Greats*. (Other movies such as *Batman* and *Space Jam* had been demonstrated earlier but were not full versions available for sale.) The Warner Brothers U.S. launch followed on March 24 but was limited to seven cities. Almost 19,000 discs were purchased in the first two weeks of the U.S. launch, more than expected. InfoTech predicted over 600 titles by the end of 1997 and more than 8,000 titles by 2000.

By December of 1997, over 1 million individual DVD discs had been shipped, representing about 530 titles. By the end of 1999, over 100 million discs had shipped, representing about 5,000 titles. A year later, over 10,000 titles were available in the United States and over 15,000 were on the market worldwide. By the end of 2001, about 14,000 titles were available in the United States. In December of 2002, about 23,000 titles were available in the United States. Compared to other launches (such as CDs and LDs), a huge number of titles had been released in a very short time. (Note that this does not include adult titles, which account for an additional 15 percent or so.) By March 2003, six years after the launch, over 1.5 billion copies of DVD titles had been shipped.

A number of DVD launches in Europe were announced with little follow-through, but DVDs became established in Europe around the end of 1998. The availability of DVD software in Europe was initially about 18 months to a year behind the United States, but this gap has shortened over the years to a delay of only a few months to a few weeks.

Many Internet databases can be used to search for DVD titles. Here are a few of the best sites:

- Internet Movie Database DVD Browser (info on all regions, us.imdb.com/Sections/DVDs)

- Doug MacLean's Home Theater Info list (region 1, downloadable list, www.hometheaterinfo.com/dvdlist.htm)

- DVD Entertainment Group (region 1, searchable and downloadable database, www.dvdinformation.com/titles)

- Widescreen Review (widescreen-specific DVD titles, www. widescreenreview.com)

- Most Internet DVD stores also have searchable lists (See "Where Can I Buy (or Rent) DVDs and Players?" in Chapter 6, which lists Web sites where you can buy or rent DVDs)

DVD-Audio started even slower than DVD-Video. The first commercially available DVD-Audio title, *Big Phat Band*, was released in October 2000 on the Silverline label of 5.1 Entertainment. Major music labels BMG Entertainment, EMI Music, Universal Music, and Warner Music have committed to DVD-Audio titles, although in the fall of 2001 Universal announced it would release SACD titles first. As of the end of 2001, just under 200 DVD-Audio titles were available. By mid-2003 about 600 DVD-Audio titles and about 900 SACD titles were available worldwide. The first SACD titles were released in Japan in May of 1999.

DVD-ROM computer software is slowly appearing. Many initial DVD-ROM titles were only available as part of a hardware or software bundle. The *International Data Corporation* (IDC) predicted that over 13 percent of all software would be available in DVD-ROM format by the end of 1998, but reality didn't meet expectations. In one sense, DVD-ROMs are simply larger, faster CD-ROMs and contain the same material. In many cases, CD-ROMs are big enough that DVD-ROMs are unnecessary, but DVD-ROMs can take advantage of the high-quality video and multichannel audio capabilities being added to many DVD-ROM-equipped computers.

Where Can I Read Reviews of DVDs?

The following sites have reviews of at least 800 discs. The list of DVD review sites at Yahoo is also recommended.

- The Big Picture (www.thebigpicturedvd.com)
- BinaryFlix (menu pictures included with every review, www.binaryflix.com)
- The Cinema Laser (www.thecinemalaser.com)
- The Digital Bits (www.thedigitalbits.com)
- DVD Authority (www.dvdauthority.com)
- DVD File (www.dvdfile.com)
- DVD Review (www.dvdreview.com)
- DVD Shrine (www.dvdshrine.com)
- DVD Talk (www.dvdtalk.com)
- DVD Verdict (www.dvdverdict.com)
- Widescreen Review Magazine (widescreen movies only, www.widescreenreview.com)

How Do I Find out when a Movie Will Be Available on DVD?

First, check one of the lists and databases mentioned in the previous section to make sure it's not already available. Then check the upcoming release lists at DVD Review and Laser Scans (www.laserscans.com/upcoming.htm). A release list is also available at Image Entertainment (www.image-enter-tainment.com), and a good source of info about unannounced titles is the Digital Bits Rumor Mill (www.thedigitalbits.com/rumormill.html).

Why Isn't My Favorite Movie on DVD?

Many factors determine when a title is released on DVD. Sometimes the director or producer has control over the DVD and video release. Other times it's up to the studio marketing group or problems may be occurring due to rights. For example, a DVD might be available in one country or region but not available in another because different studios have distribution rights in different countries. Studios do listen to customers, so let them know which titles you'd like to see (see Chapter 6's "Studios, Video Publishers, and Distributors").

How Can I Find DVDs with Specific Features or Characteristics?

Use one of the searchable databases listed previously and select the features you're looking for (anamorphic widescreen, French audio track, Flemish subtitles, and so on). If a database doesn't include the characteristics you're looking for, try a different database.

Why Do Some Rental Stores Not Carry Widescreen DVDs?

Some rental chains such as Blockbuster and retailers such as Wal-Mart originally carried only full-screen (pan and scan) versions of movies when both widescreen and full-screen versions were available. This infuriated many DVD fans, who could never countenance watching a non-widescreen version of a movie on DVD. There was much complaining, including an online petition with over 25,000 signatures. In early 2003 Blockbuster reversed their policy with the following statement: "We made a decision to purchase the majority of titles we bring in on DVD in the widescreen format. We try to follow our customer preferences. As DVD becomes increasingly popular, they become more familiar with the features and with the benefits of letterboxing. They've learned it's a superior format to full-frame." Wal-Mart similarly switched to widescreen versions apparently after realizing that they sold better.

See "What's Widescreen? How Do the Aspect Ratios Work?" in Chapter 3 for more about widescreen. See "How Do I Get Rid of the Black Bars at the Top and Bottom?" later in the chapter for the pros and cons of letterboxing.

How Much Do Players and Drives Cost?

The prices of mass-market DVD movie players range from $40 to $3,000. (Refer to the earlier "Which DVD Players and Drives Are Available?" for more information.) DVD-ROM drives and upgrade kits for computers sell for around $30 to $400. *Original Equipment Manufacturer* (OEM) drive prices are around $40.

How Much Do Discs Cost?

It varies, but most DVD movies are listed at $20 to $30, with street prices between $15 and $25, even those with supplemental material. Low-priced movies can be found for under $10. So far DVDs have not followed the initial high-rental-price model of VHS.

DVD-ROMs are usually slightly more expensive than CD-ROMs because they contain more, they cost more to replicate, and the market is smaller. But as the installed base of drives grow, DVD-ROMs will eventually cost about the same as CD-ROMs do today.

The following sites help you find the lowest prices and discount coupons:

- BargainFlix (www.bargainflix.com)
- DVD Price Search (www.dvdpricesearch.com)

How Are DVDs Doing? Where Can I Get Statistics?

DVD did not take off quite as fast as some early predictions, but it has sold faster than videotape, CD, and laserdisc. In fact, before its third birthday in March 2000, DVD had become the most successful consumer electronics entertainment product ever.

Here are some predictions:

- **InfoTech (1995)**   Worldwide sales of DVD players in 1997 will be 800,000. Worldwide sales of DVD-ROM drives in 1997 will be 1.2 million, with sales of 39 million drives in 2000.
- **C-Cube (1996)**   1 million players and drives in 1997.
- **InfoTech (1996)**   820,000 DVD-Video players in the first year, with 80 million by 2005.
- **Philips (1996)**   25 million DVD-ROM drives worldwide by 2000 (10 percent of the projected 250 million optical drives).
- **Pioneer (1996)**   500,000 DVD-ROM drives sold in 1997, with 54 million sold in 2000.
- **Pioneer (1996)**   400,000 DVD-Video players in 1996, with 11 million by 2000, and 100,000 DVD-Audio players in 1996, with 4 million by 2000.
- **Time Warner (1996)**   10 million DVD players in the United States by 2002.
- **Toshiba (1996)**   100,000 to 150,000 DVD-Video players will be sold in Japan between November 1 and December 31, 1996, and 750,000 to 1 million by November 1, 1997. (The actual count of combined shipments by Matsushita, Pioneer, and Toshiba was 70,000 in October through December of 1996.) The total worldwide DVD hardware market is expected to reach 120 million units in the year 2000. The worldwide set-top DVD player market will be 2 million units in the first year, with sales of 20 million in the year 2000.

- **Toshiba (1996)** 120 million DVD-ROM drives in 2000 (80 percent penetration of 100 million PCs). Toshiba says they will no longer make CD-ROM drives in 2000.

- **AMI (1997)** An installed base of 7 million DVD-ROM drives by 2000.

- **BASES** 3 million DVD-Video players sold in the first year, with 13 million sold in the sixth year.

- **CEMA (1997)** 400,000 DVD-Video players in United States in 1997, with 1 million in 1998.

- **Dataquest (1997)** Over 33 million shipments of DVD players and drives by 2000.

- **Forrester Research (1997)** A U.S. base of 53 million DVD-equipped PCs established by 2002 and 5.2 percent of U.S. households (5 million) will have a DVD-V player in 2002; 2 percent will have a DVD-Audio player.

- **IDC (1997)** 10 million DVD-ROM drives sold in 1997, 70 million in 2000 (surpassing CD-ROMs), and 118 million in 2001, with over 13 percent of all software available on DVD-ROM in 1998. DVD-R drives will be more than 90 percent of combined CD/DVD-R market in 2001.

- **Intel (1997)** 70 million DVD-ROM drives by 1999 (sales will surpass CD-ROM drives in 1998).

- **Paul Kagan (1997)** 800,000 DVD players in the United States in 1997, 10 million in 2000, and 40 million in 2006 (43 percent penetration), and 5.6 million discs sold in 1997, 172 million discs in 2000, and 623 million in 2006.

- **Microsoft (Peter Biddle, 1997)** 15 million DVD PCs sold in 1998, with 50 million DVD PCs sold in 1999.

- **SMD (1997)** 100 million DVD-ROM/RAM drives shipped in 2000.

- **InfoTech (January 1998)** 20 million DVD-Video players worldwide in 2002, with 58 million by 2005. 99 million DVD-ROM drives worldwide in 2005. No more than 500 DVD-ROM titles available by the end of 1998. About 80,000 DVD-ROM titles available by 2005.

- **Microsoft (Jim Taylor, 1998)** An installed base of 35 million DVD PCs in 1999.

- **Screen Digest (December 1998)** In 1998, 125,000 DVD-Video players will be in European homes, with 485,000 in 1999, and 1 million in 2000.

- **Yankee Group (January 1998)**   By 1998, 650,000 DVD-Video players will be in use, with 3.6 million by 2001 and 19 million DVD-PCs by 2001.
- **Baskerville (April 2000)**   Worldwide spending on DVD software will surpass that of VHS by 2003. A worldwide installed base of 625 million DVD players will occur by 2010 (55 percent of TV households).
- **IDC (July 2000)**   70 million DVD players and drives will be sold by year's end.
- **IRMA (April 2000)**   12 million players will ship worldwide in 2000.
- **Jon Peddie (June 2000):** Almost 20 million DVD players will be sold in the United States in 2004.
- **Japanese Electronics and Information Technologies Association (December 2000)**   37 million DVD players worldwide by 2001.
- **Screen Digest (June 2000)**   The European installed base of DVD-Video players will be 0.3 million in 1998, 1.5 millionin 1999, 5.4 million in 2000, and 47.1 million in 2003.
- **DVD Entertainment Group (July 2001)**   Approximately 30 million DVD players will be sold in the United States by the end of 2001.
- **Understanding & Solutions (April 2002)**   DVD player penetration in the United Kingdom could grow to 70 percent by 2006 (CD player penetration reached only 50 percent in the same time period after its launch).

Here's reality:

- **1997**
  - 349,000 DVD-Video players shipped in the United States (about 200,000 sold into homes)
  - 900 DVD-Video titles available in the United States, with over 5 million copies shipped and about 2 million sold
  - Over 500,000 DVD-Video players shipped worldwide
  - Around 330,000 DVD-ROM drives shipped worldwide with about 1 million bundled DVD-ROM titles
  - 60 DVD-ROM titles (mostly bundled)

- **1998**
  - 1,089,000 DVD-Video players shipped in the United States (an installed base of 1,438,000)

- 400 DVD-Video titles in Europe (135 movie and music titles)
- 3,000 DVD-Video titles in the United States (2,000 movie and music titles)
- 7.2 million DVD-Video discs purchased

- **1999**

  - 4,019,000 DVD-Video players shipped in the United States (an installed base of 5,457,000)
  - Over 6,300 DVD-Video titles in the United States
  - About 26 million DVD-ROM drives worldwide
  - About 75 DVD-ROM titles available in the United States

- **2000**

  - 8.5 million DVD-Video players shipped in the United States (an installed base of 13,922,000)
  - About 46 million DVD-ROM drives worldwide
  - Over 10,000 DVD-Video titles available in the United States
  - Belgium: An installed base of 100 thousand
  - France: An installed base of 1.2 million
  - Germany: An installed base of 1.2 million
  - Italy: An installed base of 360 thousand
  - Netherlands: An installed base of 200 thousand
  - Spain: An installed base of 300 thousand
  - Sweden: An installed base of 120 thousand
  - Switzerland: An installed base of 250 thousand
  - UK: An installed base of 1 million

- **2001**

  - 12.7 million DVD-Video players shipped in the United States (An installed base of 26,629,000)
  - Over 45 million DVD-ROM drives in the United States
  - Over 90 million DVD-ROM drives worldwide
  - UK: An installed base of 3 million

- **2002**

  - 17 million DVD-Video players shipped in the United States (an installed base of 43,718,000)
  - Over 75 million DVD-ROM drives in the United States
  - Over 140 million DVD-ROM drives worldwide

For comparison, in 1997 about 700 million audio CD players and 160 million CD-ROM drives were in use worldwide. That same year 1.2 billion CD-ROMs were shipped worldwide with about 46,000 different titles available. About 80 million VCRs were owned in the United States (89 percent of households), with about 400 million worldwide, and 110,000 VCRs shipped in the first two years after their release. Nearly 16 million VCRs were shipped in 1998. In 2000, about 270 million TVs were owned in the United States, with 1.3 billion worldwide. When DVDs came out in 1997, under 3 million laserdisc players were being used in the United States.

For the latest U.S. player sales statistics, see the CEA page at The Digital Bits. Other DVD statistics and forecasts can be found at IRMA, Media-Line, and Twice. Industry analyses and forecasts can be purchased from Adams Media Research, the British Video Association, Cahners In-stat, eBrain, IDC, Screen Digest, Understanding & Solutions, and others.

What Are Regional Codes, Country Codes, or Zone Locks?

Motion picture studios want to control the home release of movies in different countries because theater releases aren't simultaneous (a movie may come out on video in the United States when it's just hitting screens in Europe). Also, studios sell distribution rights to different foreign distributors in order to guarantee an exclusive market. Therefore, they required that the DVD standard include codes to prevent the playback of certain discs in certain geographical regions. Each player is given a code for the region in which it's sold and will refuse to play discs that are not coded for its region. This means that a disc bought in one country may not play on a player bought in another country. Some people believe that region codes are an illegal restraint of trade, but no legal cases have established this.

Regional codes are entirely optional for the disc maker to include. Discs without region locks will play on any player in any country. It's not an encryption system; it's just one byte of information on the disc that the player checks. Some studios originally announced that only their new releases would have regional codes, but so far almost all Hollywood releases play in only one region. Region codes are also a permanent part of the disc, and they won't unlock after a period of time. Region codes don't apply to DVD-Audio, DVD-ROM, or recordable DVDs.

Seven regions (also called locales) have been established and each one is assigned a number. Players and discs are often identified by the region number superimposed on a globe. If a disc plays in more than one region, it will have more than one number on the globe. The region codes are as follows:

1. U.S., Canada, and U.S. Territories
2. Japan, Europe, South Africa, and Middle East (including Egypt)
3. Southeast Asia and East Asia (including Hong Kong)
4. Australia, New Zealand, Pacific Islands, Central America, Mexico, South America, and the Caribbean
5. Eastern Europe (Russia), Indian subcontinent, Africa, North Korea, and Mongolia
6. China
7. Reserved
8. Special international venues (airplanes, cruise ships, and so on)

See the map at www.blackstar.co.uk/help/help_dvd_regions.

Technically, no such thing as a region zero disc or a region zero player exists. However, an all-region disc does exist, and all-region players are available as well. Some players can be hacked using special command sequences from the remote control to switch regions or to play all region codes. Some players can be physically modified (or chipped) to play discs regardless of the regional codes on the disc. This usually voids the warranty, but it is not illegal in most countries, because the only thing that requires player manufacturers to region-code their players is the *Content Scrambling System* (CSS) license (see "What Are the Copy Protection Issues?"). Many retailers, especially outside North America, sell players that have already been modified for multiple regions, or in some cases they simply provide instructions on how to access the "secret" region change features already built into the player.

As an interesting side note, on February 7, 2001, NASA sent two *multi-region DVD players* to the International Space Station (read about it at www.techtronics.com/uk/shop/510-nasa.html).

Extensive information about modifying players and buying region-free players can be found on the Internet (see Chapter 6's "DVD Utilities and Region-Free Information").

In addition to region codes, differences in discs for NTSC and PAL TV systems also exist (see "Is DVD-Video a Worldwide Standard? Does It Work with NTSC, PAL, and SECAM?" later in this chapter).

Some discs from Fox, Buena Vista/Touchstone/Miramax, MGM/Universal, Polygram, and Columbia TriStar contain program code that checks for the proper region setting in the player. (*There's Something About Mary* and *Psycho* are examples.) In late 2000, Warner Brothers began using the same active region-code-checking system that other studios had been using for over a year. They called it *region code enhancement* (RCE, also known as REA), and it received much publicity. RCE was first added to discs such as *The Patriot* and *Charlie's Angels*.

"Smart discs" with active region checking won't play on *code-free* players set for all regions (FFh), but they can be played on manual *code-switchable* players that enable you to use the remote control to change the player's region to match the disc. Smart discs also may not work on *auto-switching* players that recognize and match the disc region. It depends on the player's default region setting. An RCE disc has all its region flags set so that the player doesn't know which one to switch to. The disc queries the player for the region setting and aborts playback if it's the wrong one. A default player setting of region 1 can fool RCE discs from region 1. Playing a region 1 disc for a few seconds sets most auto-switching players to region 1 and thus enables them to play an RCE disc.

When an RCE disc detects the wrong region or an all-region player, it usually puts up a message saying that the player may have been altered and that the disc is not compatible with the player. A serious side effect is that some legitimate players fail the test, such as the Fisher DVDS-1000.

RCE's first appearance caused much wailing and gnashing of teeth, but DVD fans quickly learned that it only affected some players (www.dvdtalk.com/rce.html). Makers of player modification kits that didn't work with RCE soon improved their chips to get around it. For every higher wall, there is a taller ladder. See DVDTalk's *RCE FAQ* for more info and workarounds.

Region codes do not apply to DVD-Audio. In general, region codes don't apply to recordable DVDs. A DVD that you make on a PC with a DVD burner or in a home DVD video recorder will play in all regions (but don't forget NTSC versus PAL differences).

Regional codes also apply to game consoles such as PlayStation 2 and Xbox, but only for DVD-Video (movie) discs (see DVDRegionX.com for region modifications to PS2). PlayStation has a separate regional lockout scheme for games.

When it comes to DVD-ROM computers, only DVD-Video discs are affected by regional codes, not DVD-ROM discs containing computer software. Computer playback systems check for regional codes before playing movies from a CSS-protected DVD-Video (see "What Are the Copy Protection Issues?" for CSS info). Newer *RPC2* DVD-ROM drives let you change the region code several times (RPC stands for *region protection control*), but once an RPC2 drive has been changed five times, it can't be changed again unless the vendor or manufacturer resets the drive. The *Drive Info* utility can tell you if you have an RPC2 drive (it will say "This drive has region protection"). See "DVD Utilities and Region-Free Information," for more information about circumventing DVD-ROM region restrictions. Since December 31, 1999, only RPC2 drives have been manufactured.

What Are the Copy Protection Issues?

Content protection system architecture (CPSA) is the name given to the overall framework for security and access control across the entire DVD family. Developed by the *4C Entity* (Intel, IBM, Matsushita, and Toshiba) in cooperation with the *Copy Protection Technical Working Group* (CPTWG), CPSA covers encryption, watermarking, and the protection of analog and digital outputs. Many forms of content protection apply to DVD, as detailed in the following seven sections.

Analog CPS (Macrovision)

Copying to videotape (analog) can be prevented with a *Macrovision* 7.0 circuit in every DVD player. The general term is *Analog Protection System* (APS), also sometimes called copyguard. Computer video cards with composite or s-video (Y/C) output must also use APS. Macrovision adds a rapidly modulated colorburst signal (Colorstripe) along with pulses in the vertical blanking signal (AGC) to the composite video and s-video outputs. This confuses the synchronization and automatic-recording-level circuitry in 95 percent of consumer VCRs. Unfortunately, it can degrade the picture, especially with old or nonstandard equipment.

Macrovision may show up as stripes of color, distortion, rolling, a black and white picture, and dark/light cycling. Macrovision creates problems for most TV/VCR combos (see "Will I Have Problems Connecting My VCR Between My TV and DVD Player?") and some high-end equipment such as line doublers and video projectors.

Macrovision was not present on the component output of early players, but it is required on component output of newer players (AGC only, because there is no burst in a component signal). DVDs contain trigger bits that tell the player whether or not to enable Macrovision AGC, with the optional addition of two-line or four-line Colorstripe. The triggers occur about twice a second, which enables fine control over the video. The producer of the disc decides the amount of copy protection to enable and then pays Macrovision royalties accordingly (several cents per disc). Just as with videotapes, some DVDs are Macrovision-protected and some aren't. For a few Macrovision details, see STMicroelectronics' NTSC/PAL video encoder datasheets at www.st.com/stonline/books/.

Inexpensive devices can defeat Macrovision, although only a few work against the new Colorstripe feature. These devices include products such as Video Clarifier, Image Stabilizer, Color Corrector, and CopyMaster (www. videoguys.com/sima.htm). You can also build your own (http://66.40.78.100/ Services/TECH_Notes/nineteen.html). Some DVD players can be modified to turn off Macrovision output. Professional *time-base correctors* (TBCs) that regenerate line 21 also remove Macrovision. APS affects only video, not audio.

Copy Generation Management System (CGMS)

Each disc contains information specifying if the contents can be copied. This is a *serial copy generation management system* (SCMS) designed to prevent initial copies or generational copies (copies of copies). The CGMS information is embedded in the outgoing video signal. For CGMS to work, the equipment making the copy must recognize and respect the CGMS information. The analog standard (CGMS-A) encodes the data on NTSC line 21 (in the *extended data service* [XDS]) or line 20. CGMS-A is recognized by most digital camcorders and by some computer video capture cards (they will flash a message such as "recording inhibited"). Professional *time-base correctors* (TBCs) that regenerate lines 20 and 21 will remove CGMS-A information from an analog signal. The digital standard (CGMS-D) is included in DTCP and HDMI for digital connections, such as *IEEE 1394/FireWire* (www.1394ta.org). See the "Digital Copy Protection System" and "High-Bandwidth Digital Content Protection" subsections.

Content Scrambling System (CSS)

Because of the potential for perfect digital copies, paranoid movie studios added a tougher copy protection requirement to the DVD standard. CSS is a data encryption and authentication scheme intended to prevent copying video files directly from DVD-Video discs. It was developed primarily by Matsushita and Toshiba. Each CSS licensee is given a key from a master set of 400 keys stored on every CSS-encrypted disc, and a license can be revoked by removing its key from future discs. The CSS decryption algorithm exchanges keys with the drive unit to generate an encryption key that is used to obfuscate the exchange of disc keys and title keys needed to decrypt data from the disc. DVD players have CSS circuitry that decrypts the data before it's decoded and displayed, and computer DVD decoder hardware and software must include a CSS decryption module. All DVD-ROM drives have extra firmware to exchange authentication and decryption keys with the CSS module in the computer.

Since 2000, new DVD-ROM drives are required to support regional management in conjunction with CSS (refer to "What Are Regional Codes, Country Codes, and Zone Locks?" and see Chapter 4's "Can I Play DVDs on My Computer?"). DVD-Video equipment manufacturers of drives, decoder chips, decoder software, and display adapters must license CSS.

CSS licenses are free, but obtaining one is a lengthy process, so it's recommended that interested parties apply early. CSS is administered by the *DVD Copy Control Association* (DVD CCA). Near the end of May 1997, CSS licenses were finally granted for software decoding. The license is extremely restrictive in an attempt to keep the CSS algorithm and keys secret. Of course, nothing that's used on millions of players and drives worldwide could be kept secret for long. In October of 1999, the CSS algorithm was cracked and posted on the Internet, triggering endless controversies and legal battles (see "What Is DeCSS? in Chapter 4).

Content Protection for Prerecorded Media (CPPM)

Content Protection for Prerecorded Media (CPPM) is used only for DVD-Audio. It was developed to be an improvement of CSS. Keys are stored in the lead-in area, but unlike CSS no title keys are placed in the sector headers. Each volume has a 56-bit album identifier, similar to a CSS disc key, stored in the control area. Each disc contains a media key block, stored in a file on the disc. The media key block data is logically ordered in rows and columns used during the authentication process to generate a decryption key from a specific set of player keys (device keys). If the device key is revoked, the media-key-block-processing step results in an invalid key value. As with CSS, the media key block can be updated to revoke the use of compromised player keys. The authentication mechanism is the same as CSS, so no changes are required for the existing drives. A disc may contain both CSS and CPPM content if it is a hybrid DVD-Video/DVD-Audio disc.

Content Protection for Recordable Media (CPRM)

Content Protection for Recordable Media (CPRM) is a mechanism that ties a recording to the media on which it is recorded. CPRM is supported by some DVD recorders, but not by many DVD players. Each blank, recordable DVD has a unique 64-bit media ID etched in the BCA (see "What Is BCA?" in Chapter 3). When protected content is recorded on the disc, it can be encrypted with a 56-bit C2 (Cryptomeria) cipher derived from the media ID. During playback, the ID is read from the BCA and used to generate a key to decrypt the contents of the disc. If the contents of the disc are copied to other media, the ID will be absent or wrong, and the data will not be decryptable.

Digital Copy Protection System (DCPS)

In order to provide digital connections between components without allowing perfect digital copies, five *digital copy protection systems* (DCPs) were proposed to the *Consumer Electronics Association* (*CEA*). The frontrunner is *Digital Transmission Content Protection* (*DTCP*), which focuses on IEEE 1394/FireWire, but it can be applied to other protocols. The draft proposal (called 5C, for the five companies that developed it) was made by Intel, Sony, Hitachi, Matsushita, and Toshiba in February of 1998. Sony released a DTCP chip in mid-1999.

Under DTCP, devices that are digitally connected, such as a DVD player and a digital TV or a digital VCR, exchange keys and authentication certificates to establish a secure channel. The DVD player encrypts the encoded audio-video signal as it sends it to the receiving device, which must decrypt it. This keeps other connected but unauthenticated devices from stealing the signal. No encryption is needed for content that is not copy protected. Security can be "renewed" by new content (such as new discs or new broadcasts) and new devices that carry updated keys and revocation lists (to identify unauthorized or compromised devices).

A competing proposal, *extended conditional access* (XCA), from Zenith and Thomson, is similar to DTCP. However, it can work with one-way digital interfaces (such as the EIA-762 RF remodulator standard) and uses smart cards for renewable security. Other proposals have been made by MRJ Technology, NDS, and Philips. In all five proposals, content is marked with CGMS-style flags of "copy freely," "copy once," "don't copy," and sometimes "no more copies." Digital devices that do nothing more than reproduce audio and video will be able to receive all data (as long as they can authenticate that they are playback-only devices). Digital recording devices can only receive data that is marked as copyable, and they must change the flag to "don't copy" or "no more copies" if the source is marked "copy once."

DCPS in general is designed for the next generation of digital TVs, receivers, and video recorders. It requires new DVD players with digital connectors (such as those on digital video equipment). These new products began to appear in 2003. Because the encryption is done by the player, no changes are needed to existing discs.

High-Bandwidth Digital Content Protection (HDCP) and HDMI

High-Bandwidth Digital Content Protection (HDCP) is similar to DTCP, but it has been designed for digital video monitor interfaces such as *digital visual interface* (DVI). In 1998, the *Digital Display Working Group* (DDWG) was formed to create a universal interface standard between computers and displays to replace the analog VGA connection standard. The resulting DVI specification, released in April 1999, was based on Silicon Image's PanelLink technology, which at 4.95 Gbps can support 1600×1200 UXGA resolution, covering all the HDTV resolutions. Intel proposed a security component for DVI: HDCP. A new connection standard called HDMI now combines DVI and HDCP, and many new HDTV displays are likely to have both IEEE 1394 and HDMI connections.

HDCP provides authentication, encryption, and revocation. Specialized circuitry in the playback device and in the display monitor encrypts video data before it is sent over the link. When an HDMI output senses that the connected monitor does not support HDCP, it lowers the image quality of protected content. The HDCP key exchange process verifies that a receiving device is authorized to display or record video. It uses an array of 40 56-bit secret device keys and a 40-bit key selection vector, all supplied by the HDCP licensing entity. If the security of a display device is compromised, its key selection vector is placed on the revocation list. The host device has the responsibility of maintaining the revocation list, which is updated by *system renewability messages* (SRMs) carried by newer devices and by video content. Once the authority of the receiving device has been established, the video is encrypted by an *exclusive OR* (XOR) operation with a stream cipher generated from keys exchanged during the authentication process. If a display device with no decryption capability attempts to display encrypted content, it appears as random noise.

The first four forms of copy protection are optional for disc producers. Movie decryption is also optional for hardware and software playback manufacturers: A player or computer without decryption capabilities will only be able to play unencrypted movies. CPRM is handled automatically by DVD recorders, whereas DCPS and HDCP are performed by the DVD player, not by the disc developer.

These copy protection schemes are designed only to guard against casual copying (which the studios claim causes billions of dollars in lost revenue). The goal is to "keep the honest people honest." The people who developed the copy protection standards are the first to admit they can't stop well-equipped pirates.

Movie studios have promoted legislation making it illegal to defeat DVD copy protection. The result is the *World Intellectual Property Organization*

(WIPO) Copyright Treaty, the WIPO Performances and Phonograms Treaty (December 1996), and the compliant U.S. *Digital Millennium Copyright Act* (DMCA), which was passed into law in October of 1998. Software intended specifically to circumvent copy protection is now illegal in the United States as well as many other countries. A cochair of the legal group of the DVD copy protection committee stated, "in the video context, the contemplated legislation should also provide some specific assurances that certain reasonable and customary home recording practices will be permitted, in addition to providing penalties for circumvention." It's not at all clear how this might be permitted by a player or by studios that routinely set the "don't copy" flag on all their discs.

DVD-ROM drives and computers, including DVD-ROM upgrade kits, are required to support Macrovision, CGMS, and CSS. PC video cards with TV outputs that don't support Macrovision do not work with encrypted movies. Computers with IEEE 1394/FireWire connections must support the final DCPS standard in order to work with other DCPS devices. Likewise, computers with HDMI (DVI) connections must support HDCP to output DVD-Video content. Every DVD-ROM drive must include CSS circuitry to establish a secure connection to the decoder hardware or software in the computer, although CSS can only be used on DVD-Video content. Of course, because a DVD-ROM can hold any form of computer data, other encryption schemes can be implemented. See "Can I Play DVD Movies on My Computer?" in Chapter 4 for more information on DVD-ROM drives.

The *Watermarking Review Panel* (WaRP) of the CPTWG, the successor to the *Data-Hiding Subgroup* (DHSG), selected an audio watermarking system that has been accepted by the DVD Forum for DVD-Audio (see the following "What About DVD-Audio or Music DVDs?"). The original seven video watermarking proposals were merged into three: IBM/NEC, Hitachi/Pioneer/Sony, and Macrovision/Digimarc/Philips. On February 17, 1999, the first two groups combined to form the Galaxy Group and merged their technologies into a single proposal. The second group has dubbed their technology Millennium.

Watermarking permanently marks each digital audio or video frame with noise that is supposedly undetectable by human ears or eyes. Watermark signatures can be recognized by playback and recording equipment to prevent copying, even when the signal is transmitted via digital or analog connections or is subjected to video processing. Watermarking is not an encryption system, but rather it is a way to identify whether a copy of a piece of video or audio can be played. New players and software are required to support watermarking, but the DVD Forum intends to make watermarked discs compatible with existing players. Reports were made that the early watermarking technique used by Divx caused visible "raindrop" or "gun-shot" patterns, but the problem was apparently solved for later releases.

What About DVD-Audio or Music DVDs?

When DVDs were released in 1996, no DVD-Audio format was available, although the audio capabilities of DVD-Video far surpassed CDs. The DVD Forum sought additional input from the music industry before defining the DVD-Audio format. A draft standard was released by the DVD Forum's *Working Group 4* (WG4) in January of 1998, and version 0.9 was released in July. The final DVD-Audio 1.0 specification (minus copy protection) was approved in February of 1999 and released in March, but products were delayed in part by the slow process of selecting copy protection features (encryption and watermarking), with complications introduced by the *Secure Digital Music Initiative* (SDMI). The scheduled October 1999 release was further delayed until mid-2000, ostensibly because of concerns caused by the CSS crack (see "What Is DeCSS?" in Chapter 4), but also because the hardware wasn't quite ready, production tools weren't up to snuff, and the support from music labels was lackluster. Pioneer released the first DVD-Audio players (without copy protection support) in Japan in late 1999.

Matsushita released Panasonic and Technics universal DVD-Audio/DVD-Video players in July of 2000 for $700 to $1,200. Pioneer, JVC, Yamaha, and others released DVD-Audio players in the fall of 2000 and early 2001. By the end of 2000, about 50 DVD-Audio titles were available, and by the end of 2001 just under 200 DVD-Audio titles were available.

DVD-Audio is a separate format from DVD-Video. DVD-Audio discs can be designed to work in DVD-Video players, but it's possible to make a DVD-Audio disc that won't play at all in a DVD-Video player. This is because the DVD-Audio specification includes new formats and features, with content stored in a separate DVD-Audio zone on the disc (the AUDIO_TS directory) that DVD-Video players never look at. New DVD-Audio players are needed to solve this problem, or new "universal players" should be produced that can play both DVD-Video and DVD-Audio discs. Universal players are also called *video-capable audio players* (VCAPs).

A plea to producers: Universal players are rare, but you can make universal discs easily. With a small amount of effort, all DVD-Audio discs can be made to work on all DVD players by including a Dolby Digital version of the audio in the DVD-Video zone.

A plea to DVD-Audio authoring system developers: Make your software do this by default or strongly recommend this option during authoring.

DVD-Audio players (and universal players) work with existing receivers. They output PCM and Dolby Digital, and some will support the optional DTS and *direct stream digital* (DSD) formats. However, most current receivers can't decode high-definition, multichannel PCM audio (see Chapter 3's "Details of DVD-Audio and SACD," for details), and even if they could, it can't be carried on standard digital audio connections. DVD-Audio players with high-end *digital-to-analog converters* (DACs) can only be hooked up to receivers with two-channel or six-channel analog inputs, but quality is lost if the receiver converts back to digital for processing. New receivers with improved digital connections such as IEEE 1394 (FireWire) are needed to use the full digital resolution of DVD-Audio.

DVD audio is copyright protected by an *embedded signaling* or *digital watermark* feature. This uses signal-processing technology to apply a digital signature and optional encryption keys to the audio in the form of supposedly inaudible noise so that new equipment will recognize copied audio and refuse to play it. Proposals from Aris, Blue Spike, Cognicity, IBM, and Solana were evaluated by major music companies in conjunction with the 4C Entity. Aris and Solana merged to form a new company called Verance, whose *Galaxy* technology was chosen for DVD-Audio in August of 1999. (In November of 1999, Verance watermarking was also selected for SDMI.) Verance and 4C claimed that tests on the Verance watermarking method showed it was inaudible, but golden-eared listeners in later tests were able to detect the watermarking noise.

Sony and Philips have developed a competing SACD format that uses DVD discs (see Chapter 3's "Details of DVD-Audio and SACD"). Sony released version 0.9 of the SACD spec in April 1998, and the final version appeared a year later. SACD technology is available to existing Sony/Philips CD licensees at no additional cost. Most initial SACD releases have been mixed in stereo, not multichannel. SACD was originally supposed to provide "legacy" discs with two layers, one that plays in existing CD players, plus a high-density layer for DVD-Audio players, but technical difficulties kept dual-format discs from being produced until the end of 2000, and only then in small quantities. Pioneer, which released the first DVD-Audio players in Japan at the end of 1999, included SACD support in their DVD-Audio players. If other manufacturers follow suit, the entire SACD versus DVD-Audio standards debate could be moot, because DVD-Audio players would play both types of discs.

Sony released an SACD player in Japan in May of 1999 at the tear-inducing price of $5,000. The player was released in limited quantities in the United States at the end of 1999. Philips released a $7,500 player in May of 2000, and Sony shipped a $750 SACD player in Japan in mid-2000. About 40 SACD titles were available at the end of 1999, from studios such as DMP,

Mobile Fidelity Labs, Pioneer, Sony, and Telarc. Over 500 SACD titles were available by the end of 2001.

A drawback related to DVD-Audio and SACD players is that most audio receivers with six channels of analog input aren't able to provide bass management. Receivers with Dolby Digital and DTS decoders handle bass management internally, but six-channel analog inputs are usually passed straight through to the amplifier. Without full bass management on six-channel analog inputs, any audio setup that doesn't have full-range speakers for all five surround channels will not properly reproduce all the bass frequencies.

If you are interested in making the most of a DVD-Audio or SACD player, you need a receiver with six-channel, analog audio inputs. You also need five full-frequency speakers (that is, each speaker should be able to handle subwoofer frequencies) and a subwoofer, unless you have a receiver that can perform bass management on the analog inputs, or you have an outboard bass management box such as one from *Outlaw Audio* (www.outlawaudio.com).

For more on DVD-Audio, including lists of titles and player models, visit the Digital Audio Guide (www.digitalaudioguide.com).

Which Studios Are Supporting DVD?

All major movie studios and most major music labels support DVD.

When DVD players became available in early 1997, Warner and Polygram were the only major movie studios to release titles. Additional titles were available from small publishers. The other studios gradually joined the DVD camp (see Chapter 6's "Who Is Making or Supporting DVD Products?" for a full list and refer to this chapter's "Which DVD Titles Are Available?" for movie info). Dreamworks was the last significant studio to announce full DVD support. Paramount, Fox, and Dreamworks initially supported only Divx, but in the summer of 1998 they each announced support for open DVDs.

Can DVDs Record from VCRs, TVs, and So On?

The answer is yes. When DVD was originally introduced in 1997, it could only play. DVD video recorders appeared in Japan at the end of 1999 and in the rest of the world at the end of 2000. Early units were expensive: from $2,500 to $4,000. DVD recorders are still quite expensive (typically $500 to $2000 as of mid-2003), but they will eventually be as cheap as VCRs. DVD recorders are already being added to satellite and cable receivers, hard-disk video recorders, and similar boxes.

A DVD recorder basically works like a VCR. It has a tuner and A/V inputs, and it can be programmed to record shows. An important difference is that you never have to rewind or fast forward. Recordings on a disc are instantly accessible, usually from an onscreen menu. Note that DVD video recorders can't copy most DVD movie discs, which are protected.

Unfortunately, more than one recordable DVD format is available, and they don't all play together nicely. It's nothing like the old VHS versus Betamax battle, as many in the press would have you believe, but it is rather confusing. See Chapter 4's "What About Recordable DVDs: DVD-R, DVD-RAM, DVD-RW, DVD+RW, and DVD+R?" to get more confused.

Don't be further confused by DVD recordable drives (DVD burners) for computers. These recorders can store data, but creating full-featured DVD-Videos requires additional software to do video encoding (MPEG), audio encoding (Dolby Digital, MPEG, or PCM), navigation and control data generation, and so on (see Chapter 5, "DVD Production").

What Happens If I Scratch the Disc? Aren't Discs Too Fragile to Be Rented?

Scratches may cause minor data errors that are easily corrected. That is, data is stored on DVDs using powerful error-correction techniques that can recover from even large scratches with no loss of data. A common misperception is that a scratch will be worse on a DVD than on a CD because of higher storage density and because video is heavily compressed. DVD data density is physically four times that of CD-ROM, so it's true that a scratch will affect more data, but DVD error correction is at least 10 times better than CD-ROM error correction and more than makes up for the density increase. It's also important to realize that MPEG-2 and Dolby Digital compression are partly based on the removal or reduction of imperceptible information, so decompression doesn't expand the data as much as might be assumed.

Major scratches may cause uncorrectable errors that will produce an *input/output* (I/O) error on a computer or show up as a momentary glitch in the DVD-Video picture. Paradoxically, sometimes the smallest scratches can cause the worst errors (because of the particular orientation and refraction of the scratch). Many schemes can conceal errors in MPEG video, which may be used in future players. See the later section "How Should I Clean and Care for DVDs?" for more information.

The industry's DVD computer advisory group specifically requested no mandatory caddies or other protective carriers. Consider that laserdiscs, music CDs, and CD-ROMs are likewise subject to scratches, but many video stores and libraries rent them. Most reports of rental disc performance are positive, although if you have problems playing a rental disc, check for scratches.

VHS Is Good Enough. Why Should I Care About DVD?

The primary advantages of DVD are video quality, surround sound, and extra features (refer to "What Are the Features of DVD-Video?"). DVD will not degrade with age or after many playings like videotape will (which is an advantage for parents with kids who watch Disney videos twice a week). This is the collectability factor present with CDs versus cassette tapes. If none of this matters to you, then VHS probably is good enough.

Is the Packaging Different from CDs?

Manufacturers were worried that customers would assume DVDs would play in their CD players, so they wanted the packaging to be different. Most DVD packages are as wide as a CD jewel box (about 5 ⅝ inches) and as tall as a VHS cassette box (about 7 ⅜ inches), as recommended by the *Video Software Dealers Association* (VSDA). However, no one is being forced to use a larger package size. Some companies use standard jewel cases or paper and vinyl sleeves. Divx discs come in paperboard and plastic Q-Pack cases the same size as a CD jewel case.

Most movies are packaged in the Amaray keep case, an all-plastic clamshell with clear vinyl pockets for inserts that's popular among consumers. Time Warner's snapper, a paperboard case with a plastic lip, is less popular. The super jewel box, the stretch limo version of a CD jewel case, is common in Europe.

What's a Dual-Layer Disc? Will It Work in All Players?

A dual-layer disc has two layers of data, one of them semitransparent so that the laser can focus through it and read the second layer. Because both layers are read from the same side, a dual-layer disc can hold almost twice as much as a single-layer disc, typically four hours of video (see "What Are the Sizes and Capacities of DVDs?" in Chapter 3 for more details). Many discs use dual layers. Initially, only a few replication plants could make dual-layer discs, but most plants now have the capability.

The second layer can use either a *parallel track path* (PTP) layout where both tracks run in parallel (for independent data or special switching effects) or an *opposite track path* (OTP) layout where the second track runs in an opposite spiral. That is, the pickup head reads out from the center on the first track and then in from the outside on the second track. The OTP layout is designed to provide continuous video across both layers. OTP is also called *reverse-spiral dual layer* (RSDL).

The layer change can occur anywhere in the video; it doesn't have to be at a chapter point. No guarantee exists that the switch between layers will be seamless. The layer change is invisible on some players, but it can cause the video to freeze for a fraction of a second or up to four seconds on other players. The seamlessness depends as much on the way the disc is prepared as on the design of the player. The advantage of two layers is that long movies can use higher data rates for better quality than a single layer. See "What Is a Layer Change? Where Is It on Specific Discs?" for more about layer changes.

Dual-layer discs can be recognized by three characteristics: 1) the gold color, 2) a menu on the disc for selecting the widescreen or letterbox version, and 3) two serial numbers on one side.

The DVD specification requires that players and drives read dual-layer discs. Few units have problems with dual-layer discs; this is a design flaw and should be corrected for free by the manufacturer. Some discs are designed with a seamless layer change that technically goes beyond what the DVD spec allows. This causes problems on a few older players.

All players and drives also play double-sided discs if you flip them over. No manufacturer has announced a model that can play both sides without someone manually flipping the disc, other than a few DVD jukeboxes. The added cost of this capability is hard to justify because discs can hold over four hours of video on one side by using two layers. (Early discs used two sides because dual-layer production was not widely supported. This is no longer a problem.) Pioneer laserdisc/DVD players can play both sides of an laserdisc, but not a DVD. (See "Will High-Definition DVDs or 720p DVDs Make Current Players and Discs Obsolete?" in Chapter 2 for information on reading both sides simultaneously.)

Is DVD-Video a Worldwide Standard? Does It Work with NTSC, PAL, and SECAM?

The MPEG video on a DVD is stored in digital format, but it's formatted for one of two mutually incompatible television systems: 525/60 (NTSC) or 625/50 (PAL/SECAM). Therefore, two kinds of DVDs exist: NTSC DVDs and PAL DVDs. Some players only play NTSC discs; others play PAL and NTSC discs, depending on which region the owner lives in (refer to "What Are Regional Codes, Country Codes, or Zone Locks?").

Almost all DVD players sold in PAL countries play both kinds of discs. These *multistandard* players partially convert NTSC to a 60 Hz PAL (4.43

NTSC) signal. The player uses the PAL 4.43 MHz color subcarrier encoding format but keeps the 525/60 NTSC scanning rate. Most modern PAL TVs can handle this pseudo-PAL signal. A few multistandard PAL players output true 3.58 NTSC from NTSC discs, which requires an NTSC TV or a multistandard TV. Some players have a switch to choose 60 Hz PAL or true NTSC output when playing NTSC discs. A few *standards-converting* PAL players convert output from an NTSC disc to standard PAL for older PAL TVs. Proper "on-the-fly" standards conversion requires expensive hardware to handle scaling, temporal conversion, and object motion analysis. Because the quality of conversion in DVD players is poor, using 60 Hz PAL output with a compatible TV provides a better picture than converting from NTSC to PAL. (Sound is not affected by video conversion.) The latest software tools such as Adobe *After Effects* and Canopus *ProCoder* do quite a good job of converting between PAL and NTSC at low cost, but they are only appropriate for the production environment (converting the video before it is encoded and put on the DVD).

Most NTSC players can't play PAL discs. A small number of NTSC players (such as Apex and SMC) can convert PAL to NTSC. External converter boxes are also available, such as the Emerson EVC1595 (at $350). High-quality converters are available from companies such as TenLab (www.tenlab.com) and Snell and Wilcox (www.snellwilcox.com). Many standards-converting players can't convert anamorphic widescreen video for 4:3 displays. See "Why Is the Picture Squished, Making Things Look Too Skinny?"

Three differences exist between discs intended for playback on different TV systems: picture dimensions and pixel aspect ratios (720×480 versus 720×576), the display frame rate (29.97 versus 25), and surround audio options (Dolby Digital versus MPEG audio). (See "What Are the Video Details?" and "What Are the Audio Details?" in Chapter 3 for details.) Video from film is usually encoded at 24 frames per second but is preformatted for one of the two required display rates. Movies formatted for PAL display are usually sped up by 4 percent at playback, so the audio must be adjusted accordingly before being encoded.

All PAL DVD players can play Dolby Digital audio tracks, but not all NTSC players can play MPEG audio tracks. PAL and SECAM share the same scanning format, so discs are the same for both systems. The only difference is that SECAM players output the color signal in the format required by SECAM TVs. Note that modern TVs in most SECAM countries can also read PAL signals, so you can use a player that only has PAL output. The only case in which you would need a player with SECAM output is for older SECAM-only TVs (and you'll probably need a SECAM RF connection). (See "What Are the Outputs of a DVD Player?" in Chapter 3.)

A producer can choose to put 525/60 NTSC video on one side of the disc and 625/50 PAL on the other. Most studios put Dolby Digital audio tracks on their PAL discs instead of MPEG audio tracks. Because of PAL's higher resolution, the movie usually takes more space on the disc than the NTSC version. See Chapter 3 for more details.

Actually, three types of DVD players exist if you count computers. Most DVD PC software and hardware can play both NTSC and PAL video, as well as both Dolby Digital and MPEG audio. Some PCs can only display the converted video on the computer monitor, but others can output it as a video signal for a TV.

The bottom line is that NTSC discs (with Dolby Digital audio) play on over 95 percent of DVD systems worldwide. PAL discs play on very few players outside of PAL countries, irrespective of regions.

What About Animation on DVD? Doesn't It Compress Poorly?

Some people claim that animation, especially hand-drawn cell animation such as cartoons and anime, does not compress well with MPEG-2 or even ends up larger than the original. Other people claim that animation is simple so it compresses better. Neither is true.

Supposedly the "jitter" between frames caused by differences in the drawings or in their alignment causes problems. An animation expert at Disney pointed out that this doesn't happen with modern animation techniques. And even if it did, the motion estimation feature of MPEG-2 would compensate for it.

Because of the way MPEG-2 breaks a picture into blocks and transforms them into frequency information, it can have a problem with the sharp edges common in animation. This loss of high-frequency information can show up as "ringing" or blurry spots along edges (called the Gibbs effect). However, at the data rates commonly used for DVD, this problem does not usually occur.

Why Do Some Discs Require Side Flipping? Can't DVDs Hold Four Hours per Side?

Even though DVD's dual-layer technology (see "What Are the Sizes and Capacities of DVDs?" in Chapter 3) enables over four hours of continuous playback from a single side, some movies are split over two sides of a disc, requiring that the disc be flipped partway through. Most "flipper" discs are made because producers are too lazy to optimize the compression or make a dual-layer disc. Better picture quality is a cheap excuse for increasing the data rate; in many cases, the video will look better if carefully encoded at a lower bit rate.

A lack of dual-layer production capability is also a lame excuse; in 1997, very few DVD plants could make dual-layer discs, but this is no longer the case. Very few players can automatically switch sides, but it's not needed because movies less than four hours long can easily fit on one dual-layer (RSDL) side.

The Film Vault at DVD Review includes a list of flipper discs. Note that a flipper is not the same as a disc with a widescreen version on one side and a pan and scan version or supplements on the other.

Why Is the Picture Squished, Making Things Look Too Skinny?

Answer: RTFM. You are watching an anamorphic picture intended for display only on a widescreen TV. (See "What's Widescreen? How Do the Aspect Ratios Work?" in Chapter 3 for technical details). You need to go into the player's setup menu and tell it your TV is standard 4:3 TV, not widescreen 16:9. It will then automatically letterbox the picture so you can see the full width at the proper proportions.

In some cases, you can change the aspect ratio as the disc is playing (by pressing the "aspect" button on the remote control). On most players you have to stop the disc before you can change the aspect. Some discs are labeled with widescreen on one side and standard on the other. In order to watch the full-screen version, you must flip the disc over. (See "How Do I Get Rid of the Black Bars at the Top and Bottom?" for more information on letterboxing.)

Apparently, most players that convert from NTSC to PAL or vice versa can't simultaneously letterbox (or pan and scan) an anamorphic picture (see "Is DVD-Video a Worldwide Standard? Does It Work with NTSC, PAL, and SECAM?"). The solutions would be to use a widescreen TV, a multistandard TV, or an external converter. Or get a better player.

Do All Videos Use Dolby Digital (AC-3)? Do They All Have 5.1 Channels?

Most DVD-Video discs contain Dolby Digital soundtracks, but some discs, especially those containing only audio, have PCM tracks. It's also possible for a 625/50 (PAL) disc to contain only MPEG audio, which is not widely used. Discs with DTS audio are required to also include a Dolby Digital audio track (in a few rare cases they have a PCM track). See "What's the Deal with DTS and DVDs?" for more on DTS.

Don't assume that the Dolby Digital label is a guarantee of 5.1 channels. A Dolby Digital soundtrack can be mono, dual mono, stereo, Dolby Surround stereo, and so on. For example, *Blazing Saddles* and *Caddyshack* have monophonic soundtracks, so the Dolby Digital soundtrack on these DVDs has only one channel. Some DVD packaging has small lettering or icons under the Dolby Digital logo that indicates the channel configuration. In some cases, a soundtrack has more than one Dolby Digital version: a 5.1-channel track and a track specially remixed for stereo Dolby Surround. It's perfectly normal for your DVD player to indicate the playback of a Dolby Digital audio track while your receiver indicates Dolby Surround. This means the disc contains a two-channel Dolby Surround signal encoded in Dolby Digital format. See Chapter 3 for more audio details.

Can DVDs Have Laser Rot?

Laserdiscs are subject to what is commonly called *laser rot*, the deterioration of the aluminum layer due to oxidation or other chemical changes. This often results from the use of insufficiently pure metal for the reflective coating created during replication, but it can be exacerbated by mechanical shear stress due to bending, warping, or thermal cycles. The large size of laserdiscs makes them flexible, so that movement along the bond between layers can break the seal, which is called *delamination*. Deterioration of the data layer can be caused by chemical contaminants or gasses in the glue, or by moisture that penetrates the plastic substrate.

Like laserdiscs, DVDs are made of two platters glued together, but DVDs are more rigid and use newer adhesives. DVDs are molded from polycarbonate, which absorbs about 10 times less moisture than the slightly hygroscopic acrylic (PMMA) used for laserdiscs.

DVDs can have delamination problems, partly because some cases or players hold too tightly to the inner hub of the disc. Delamination by itself can cause problems (because the data layer is no longer at the correct distance from the surface) and can also lead to oxidation.

So far DVDs have had few "DVD rot" problems. Reports have been made of a few discs going bad, possibly due to delamination, contaminated adhesive, chemical reactions, or oxidation of the reflective layer (see www. mindspring.com/~yerington/ and www.andraste.org/discfault/discfault. htm). Occurrences of "cloudiness" or "milkiness" in DVDs have been reported, a possible cause being improper replication. An example of this would be when the molten plastic cools off too fast or isn't under enough pressure to completely fill all the bumps in the mold (see www. tapediscbusiness.com/issues/1998/0998/cloud.htm). Minimal clouding doesn't hurt playback and doesn't seem to deteriorate. If you can see something with your naked eye, it is probably not oxidation or other deterioration.

The result of deterioration is that a disc that played perfectly when it was new develops problems later, such as skipping, freezing, or picture breakup. If a disc seems to go bad, make sure it's not dirty, scratched, or warped (see "How Should I Clean and Care for DVDs?"). Try cleaning it and try playing it in other players. If the disc consistently has problems, it may have deteriorated. If so, you can't do anything to fix it. Request a replacement from the supplier.

Which Titles Are Pan and Scan Only? Why?

Some titles are available only in pan and scan because no letterbox or anamorphic transfer was made from film. (See Chapter 3 for more info on pan and scan and anamorphic formats.) Because transfers cost $50,000 to $100,000, studios may not think a new transfer is justified. In some cases, the original film or rights to it are no longer available for a new transfer. In the case of old movies, they were shot full frame in the 1.37 "academy" aspect ratio, so no widescreen version can be created. Videos shot with TV cameras, such as music concerts, are already in 4:3 format.

How Do I Make the Subtitles on My Pioneer Player Go Away?

On the remote control, press "subtitle" and then either press "clear" or 0 (zero). You have no need to use the menus.

What Is a Layer Change? Where Is It on Specific Discs?

Some movies, especially those over two hours long or encoded at a high data rate, are spread across two layers on one side of the disc. When the player changes to the second layer, the video and audio may freeze for a moment. The length of the pause depends on the player and on the layout of the disc. The pause is not a defect in the player or the disc. See "What's a Dual-Layer Disc? Will It Work in All Players?" for details.

A list of layer switch points can be found at the Film Vault of DVD Review. Please send new times to info@dvdreview.com.

The Disc Says Dolby Digital. Why Do I Get Two-Channel Surround Audio?

Some discs (many from Columbia TriStar) have two-channel Dolby Surround audio (or plain stereo) on track one and 5.1-channel audio on track two. Because some studios create separate sound mixes optimized for Dolby Surround or stereo, they feel the default track should match the majority of sound systems in use. Unless you specifically select the 5.1-channel track (using the audio button on the remote or the onscreen menu), the player will play the default two-channel track. (Some players have a feature to automatically select the first 5.1 track.) Dolby Digital doesn't necessarily mean 5.1 channels. (Refer to "Do All Videos Use Dolby Digital (AC-3)? Do They All Have 5.1 Channels?" and see Chapter 3's "What Are the Audio Details?")

Why Doesn't the Repeat A—B Feature Work on Some Discs?

Almost all features of DVDs, such as search, pause, and scan, can be disabled by the disc, which can prevent the player from backing up and repeating a segment. If the player uses a time search to repeat a segment, a disc with fancy nonsequential title organization may also block the repeat feature. In many cases, the authors don't even realize they have prevented the use of this feature.

What's the Difference Between First-, Second-, and Third-Generation DVDs?

This question has no absolute answer, because you'll get a different response from everyone you ask. The terms second generation, third generation, and so on refer both to DVD-Video players and DVD-ROM drives. In general, they simply mean newer versions of DVD playback devices. The terms haven't been used (yet) to refer to DVD products that can record, play video games, and so on.

According to some people, second-generation DVD players came out in the fall of 1997, and third-generation players were released at the beginning of 1998. According to others, the second generation of DVDs will be HD players (see Chapter 2's "Will High-Definition DVDs or 720p DVDs Make Current Players and Discs Obsolete?") that won't come out until 2003 or so. Many conflicting variations occur between these extremes, including the viewpoint that DTS-compatible players, Divx players, progressive-scan players, 10-bit video players, or players that can play *The Matrix* constitute the second, third, or fourth generation.

Things are a little more clear cut on the PC side, where second generation (DVD II) usually means 2x DVD-ROM drives that can read CD-Rs, and third generation (DVD III) usually means 5x (or sometimes 2x or 4.8x or 6x) DVD-ROM drives, a few of which can read DVD-RAMs, and some of which are RPC2 format. Some people refer to RPC2 drives or 10x drives as fourth generation. (See "What Are the Features and Speeds of DVD-ROM Drives?" in Chapter 4 for more speed info. Refer to "What Are Regional Codes, Country Codes, or Zone Locks?" for an RPC2 explanation.)

What's a Hybrid DVD?

Do you really want the answer to this one? Okay, you asked for it . . .

- It's a disc that works in both DVD-Video players and DVD-ROM PCs. (More accurately called an *enhanced* DVD.)

- It's a DVD-ROM disc that runs on Windows and Mac OS computers (more accurately called a *cross-platform* DVD).

- It's a DVD-ROM or DVD-Video disc that also contains Web content for connecting to the Internet (more accurately called a *WebDVD* or *enhanced* DVD).

- It's a disc that contains both DVD-Video and DVD-Audio content (more accurately called a *universal* or *AV* DVD.)

- It's a disc with two layers, one that can be read in DVD players and one that can be read in CD players (more accurately called a *legacy* or *CD-compatible* disc). At least three variations of this hybrid exist (although not all are commercially available).

  - It's a 0.9 to 1.2 millimeter CD substrate bonded to the back of a 0.6 millimeter DVD substrate. One side can be read by CD players, the other side by DVD players. The resulting disc is 0.6 millimeters thicker than a standard CD or DVD, which can cause problems in players with tight tolerances, such as portables. Sonopress, the first company to announce this type, calls it DVDPlus. It's colloquially known as a fat disc. It contains a variation in which an 8-centimeter data area is embedded in a 12-centimeter substrate so that a label can be printed on the outer ring.

  - It's a 0.6-millimeter CD substrate bonded to a semitransparent 0.6-millimeter DVD substrate. Both layers are read from the same side, with the CD player being required to read through the semi-transparent DVD layer, causing problems with some CD players used by SACO.

- It's a 0.6-millimeter CD substrate, with a special refractive coating that causes a 1.2-millimeter focal depth, bonded to the back of a 0.6-millimeter DVD substrate. One side can be read by CD players, the other side by DVD players.

- It's a disc with two layers or two sections, one containing pressed (DVD-ROM) data and one containing rewritable (DVD-RAM and so on) media for recording and rerecording (more accurately called a *DVD-PROM, mixed-media,* or *rewritable sandwich* disc.).

- It's a disc with two layers on one side and one layer on the other (more accurately called a *DVD-14*).

- It's a disc with an embedded memory chip for storing custom usage data and access codes (more accurately called a *chipped* DVD).

What's the Deal with DTS and DVD?

DTS Digital Surround is an audio encoding format similar to Dolby Digital. It requires a decoder, either in the player or in an external receiver (see Chapter 3's "Audio Details of DVD-Video" for technical details). Some people claim that, because of its lower compression level, DTS sounds better than Dolby Digital. Others claim no perceptible difference can be discerned, especially at the typical data rate of 768 Kbps, which is 60 percent more than Dolby Digital. Because of the many variances in production, mixing, decoding, and reference levels, it's almost impossible to accurately compare the two formats (DTS usually produces a higher volume level, causing it to sound better in casual comparisons).

DTS originally did all encoding in house, but as of October 1999 DTS encoders became available for purchase. DTS titles are generally considered to be specialty items intended for audio enthusiasts. So some DTS titles are also available in a Dolby Digital-only version.

DTS is an optional format on DVD. Contrary to uninformed claims, the DVD specification has included an ID code for DTS since 1996 (before the spec was even finalized). Because DTS was slow in releasing encoders and test discs, players made before mid-1998 (and many since) ignore DTS tracks. A few demo discs were created in 1997 by embedding DTS data into a PCM track (the same technique used with CDs and laserdiscs), and these are the only DTS DVD discs that work on all players. New DTS-compatible players arrived in mid-1998, but theatrical DTS discs using the proper DTS audio stream ID did not appear until January 7, 1999 (they were originally scheduled to arrive in time for Christmas 1997). *Mulan*, a direct-to-video animation (not the Disney movie) with DTS soundtrack, appeared in November 1998. DTS-compatible players carry an official DTS Digital Out logo.

Dolby Digital or PCM audio is required on 525/60 (NTSC) discs, and because both PCM and DTS together don't usually leave enough room for quality video encoding of a full-length movie, essentially every disc with a DTS soundtrack also carries a Dolby Digital soundtrack. This means that all DTS discs work in all DVD players, but a DTS-compatible player and a DTS decoder are required to play the DTS soundtrack. DTS audio CDs work on all DVD players, because the DTS data is encapsulated into standard PCM tracks that are passed untouched to the digital audio output. DTS discs often carry a Dolby Digital 2.0 track in Dolby Surround format instead of a full Dolby Digital 5.1 track.

Why Is the Picture Black and White?

You are probably trying to play an NTSC disc in a PAL player, but your PAL TV is not able to handle the signal. If your player has a switch or onscreen setting to select the output format for NTSC discs, choosing PAL (60 Hz) may solve the problem. (Refer to "Is DVD-Video a Worldwide Standard? Does It Work with NTSC, PAL, and SECAM?" for more information.)

You may have connected one of the component outputs (Y, R-Y, or B-Y) of your DVD player to the composite input of your TV. (See Chapter 3 for hookup details.)

Why Are Both Sides Full-Screen when One Side Is Supposed to Be Widescreen?

Many DVDs are labeled as having widescreen (16:9) format video on one side and standard (4:3) on the other. If you think both sides are the same, you're probably seeing uncompressed 16:9 on the widescreen side. It may look like 4:3 pan and scan, but if you look carefully you'll discover that the picture is horizontally compressed. The problem is that your player has been set for a widescreen TV. Refer to "Why Is the Picture Squished, Making Things Look Too Skinny?" for details.

Why Are the Audio and Video out of Sync?

There have been numerous reports of lip sync problems, where the audio lags slightly behind the video, or sometimes precedes the video. Perception of a sync problem is highly subjective; some people are bothered by it, whereas others can't discern it at all. Problems have been reported on a variety of players, notably the Pioneer 414 and 717 models (possibly all Pioneer models), some Sony models (including the 500 series and the PS2), some Toshiba models (including the 3109), and some PC decoder cards. Certain discs are also more problematic, such as *Lock, Stock, and Two Smoking Barrels, Lost In Space, Tron, The Parent Trap,* and *Austin Powers.*

The cause of the sync problem is a complex interaction of as many as four factors:

- Improper sync in audio/video encoding or DVD-Video formatting
- Poor sync during film production or editing (especially postdubbing or looping)
- Loose sync tolerances in the player
- Delay in the external decoder/receiver

The first two factors usually must be present in order for the third or fourth ones to become apparent. Some discs with severe sync problems have been reissued after being reencoded to fix the problem. In some cases, the sync problem in players can be fixed by pausing or stopping playback and then restarting, or by turning the player off, waiting a few seconds, and then turning it back on.

A good way to test your player is to simultaneously listen to the analog and digital outputs (play the digital output through your stereo and the analog output through your TV). If the audio echoes or sounds hollow, the player is delaying the signal and is thus the main cause of the sync problem.

Unfortunately, this sync problem has no simple answer and no simple fix. More complaints from customers should motivate manufacturers to take the problem more seriously and correct it in future players or with firmware upgrades. Pioneer originally stated that altering the audio-visual synchronization of their players "to compensate for the software quality would dramatically compromise the picture performance." Since then Pioneer has fixed the problem on its new players. If you have an older model, check with Pioneer about an upgrade. For more details, see Michael D.'s Pioneer Audio Sync page (www.michaeldvd.com.au/DV505/PioneerAudio).

Why Does the Picture Alternate Between Light and Dark?

You are seeing the effects of Macrovision copy protection (refer to "What Are the Copy Protection Issues?"). This is probably because you are running your DVD player through your VCR or VCR/TV combo (see Chapter 3's

"Will I Have Problems Connecting My VCR Between My TV and My DVD Player?").

How Do I Find Easter Eggs and Other Hidden Features?

Some DVD movies contain hidden features, often called Easter eggs. These are extra screens or video clips hidden in the disc by the developers. For example, *Dark City* includes scenes from *Lost in Space* and from the *Twin Peaks* movie buried in the biography pages of William Hurt and Keifer Sutherland. An amusing Shell Beach game is also entwined throughout the menus. On *Mallrats*, perhaps indicating that DVD has already become too postmodern for its own good, a hidden clip of the director telling you to stop looking for Easter eggs and do something useful is included. It's more fun to search for hidden features on your own, but if you need some help, the best list is at the DVD Review web site.

How Do I Get Rid of the Black Bars at the Top and Bottom?

The black bars are part of the *letterbox* process (see video details in Chapter 3), and in many cases you can't get rid of them. If you set the display option in your player to pan and scan (sometimes called full screen or 4:3) instead of letterbox, it won't do you much good because almost no DVD movies have been released with this feature enabled. If you set the player to 16:9 widescreen output, it will make the bars smaller, but you will get a tall, stretched picture on a standard TV.

In some cases, both a full-screen and a letterbox version of the movie may be included on the same disc, with a variety of ways to get to the full-screen version (usually only one works, so you may have to try all three):

- Check the other side of the disc (if it's two-sided).
- Look for a full-screen choice in the main menu.
- Use the "aspect" button on the remote control.

DVD was designed to make movies look as good as possible on TV. Because most movies are wider than most TVs, letterboxing preserves the format of the theatrical presentation. (Nobody seems to complain about letterboxing in theaters.) DVD is ready for TVs of the future, which are widescreen. For these and other reasons, many movies on DVD are only available in widescreen format.

About two-thirds of widescreen movies are filmed at the 1.85 (flat) aspect ratio or less. In this case, the actual size of the image on your TV is the same for a letterbox version and a full-frame version, unless the pan and scan technique is used to zoom in (which cuts off part of the picture). In other words, the picture is the same size, with extra areas visible at the top and bottom in the full-screen version. Put yet another way, letterboxing covers over the part of the picture that was also covered in the theater. Letterboxing could also allow the entire widescreen picture to be visible for movies wider than 1.85, in which case the letterboxed picture is smaller and has less detail than a pan and scan version would.

If no full-screen version of the movie is included on the disc, one solution is to use a DVD player with a zoom feature to enlarge the picture enough to fill the screen. This will cut off the sides of the picture, but in many cases it's a similar effect to the pan and scan process. Just think of it as "do-it-yourself pan and scan."

For a detailed explanation of why most movie fans prefer letterboxing, see the Letterbox/Widescreen Advocacy Page (www.widescreen.org). For an explanation of anamorphic widescreen and links to more information and examples on other web sites, see "What's Widescreen? How Do the Aspect Ratios Work?" in Chapter 3. The best solution to this entire mess might be the FlikFX Recomposition System, "the greatest advance in entertainment in 57 years" (www.widescreenmuseum.com/flikfx).

How Should I Clean and Care for DVDs?

Because DVDs are read by a laser, they are resistant to fingerprints, dust, smudges, and scratches—to a point (refer to "What Happens If I Scratch the Disc? Aren't Discs Too Fragile to Be Rented?"). However, surface contaminants and scratches can cause data errors. On a video player, the effect of data errors ranges from minor video artifacts to frame skipping to complete unplayability. So it's a good idea to take care of your discs. In general, treat them the same way you would a CD.

Your player can't be harmed by a scratched or dirty disc unless globs of nasty substances on it might actually touch the lens. Still, it's best to keep your discs clean, which will also keep the inside of your player clean. Never attempt to play a cracked disc, as it could shatter and damage the player. It doesn't hurt to leave the disc in the player, even if it's paused and still spinning, but leaving it running unattended for days on end might not be a good idea.

In general, cleaning the lens on your player is unnecessary, because the air moved by the rotating disc keeps it clean. However, if you use a lens cleaning disc in your CD player, you may wish to to the same with your DVD player. It's advisable to use a cleaning disc designed for DVD players, because minor differences exist between the position of the lens in DVD and CD players.

concerns. Panasonic also released a progressive-scan player (DVD-H1000, $3000) in the fall of 1999. Many manufacturers have released progressive models since then at progressively cheaper prices (pun intended). It's also possible to buy an external *line multiplier* to convert the output of a standard DVD player to progressive scanning. All DVD computers are progressive players, because computer monitors are progressive-scan, but quality varies. (See Chapter 4's "Can I Play DVD Movies on My Computer?" and Chapter 2's "Will High-Definition DVDs or 720p DVDs Make Current Players and Discs Obsolete?")

Converting interlaced DVD-Video to progressive video involves much more than putting film frames back together. This conversion can be done in essentially four ways:

- **Reinterleaving (also called weave)** If the original video is from a progressive source, such as film, the two fields can be recombined into a single frame.

- **Line doubling (also called bob)** If the original video is from an interlaced source, simply combining two fields will cause motion artifacts (the effect is reminiscent of a zipper), so each line of a single field is repeated twice to form a frame. The best line doublers use *interpolation* to produce new lines that are a combination of the lines above and below. The term line doubler is vague, because cheap line doublers only bob, whereas expensive line doublers (those that contain digital signal processors) can also weave.

- **Field-adaptive deinterlacing** This examines individual pixels across three or more fields and selectively weaves or bobs regions of the picture as appropriate. These chips used to cost $10,000 and up, but the feature is now appearing in consumer DVD players.

- **Motion-adaptive deinterlacing** This examines MPEG-2 motion vectors or does massive image processing to identify moving objects in order to selectively weave or bob regions of the picture as appropriate. Most systems that do this well cost $50,000 and up (aside from the cool but defunct Chromatic Mpact2 chip).

Three common kinds of deinterlacing systems are available:

- **Integrated** This is usually the best, where the deinterlacer is integrated with the MPEG-2 decoder so that it can read MPEG-2 flags and analyze the encoded video to determine when to bob and weave. Most DVD computers use this method.

- **Internal** The digital video from the MPEG-2 decoder is passed to a separate deinterlacing chip. The disadvantage is that MPEG-2 flags and motion vectors may no longer be available to help the deinterlacer determine the original format and cadence. (Some internal chips receive the repeat_first_field and top_field_first flags passed from the decoder, but not the progressive_scan flag.)

- **External**   Analog video from the DVD player is passed to a separate deinterlacer (line multiplier) or to a display with a built-in deinterlacer. In this case, the video quality is slightly degraded from being converted to analog, back to digital, and often back again to analog. However, for high-end projection systems, a separate line multiplier (which scales the video and interpolates to a variety of scanning rates) may achieve the best results.

Most progressive DVD players use an internal deinterlacing chip usually from Genesis/Faroudja. The Princeton PVD-5000 uses a Sigma Designs decoder with integrated deinterlacing, while the JVC XV-D723GD uses a custom decoder with integrated deinterlacing. Toshiba's "Super Digital Progressive" players and the Panasonic HD-1000 use 4:4:4 chroma oversampling, which provides a slight quality boost from DVD's native 4:2:0 format. Add-on internal deinterlacers such as the Cinematrix and MSB Progressive Plus are available to convert existing players to progressive-scan output. Faroudja, Silicon Image (DVDO), and Videon (Omega) line multipliers are examples of external deinterlacers.

A progressive DVD player has to determine whether the video should be line-doubled or reinterleaved. When reinterleaving film-source video, the player also has to deal with the difference between the film frame rate (24 Hz) and the TV frame rate (30 Hz). Because the 2-3 pulldown trick can't be used to spread film frames across video fields, worse motion artifacts occur than with interleaved video. However, the increase in resolvable detail more than makes up for it. Advanced progressive players such as the Princeton PVD-5000 and DVD computers can get around the problem by displaying at multiples of 24 Hz, such as 72, 96, and so on.

A progressive player also has to deal with problems such as video that doesn't have clean cadence (such as when it's edited after being converted to interlaced video, when bad fields are removed during encoding, when the video is speed-shifted to match the audio track, and so on). Another problem is that many DVDs are encoded with incorrect MPEG-2 flags, so the reinterleaver has to recognize and deal with pathological cases. In some instances, it's practically impossible to determine if a sequence is 30-frame interlaced video or 30-frame progressive video. For example, the documentary on *Apollo 13* is interlaced-video-encoded as if it were progressive. Other examples of improper encoding are *Titanic*, *Austin Powers*, *Fargo*, *More Tales of the City*, the *Galaxy Quest* theatrical trailer, and *The Big Lebowski* making-of featurette.

One problem is that many TVs with progressive input don't allow the aspect ratio to be changed. They are programmed to interpret all progressive-scan input as anamorphic. When a nonanamorphic (4:3) picture is sent to these TVs, they distort it by stretching it out. Before you buy a DTV, make sure that it enables aspect ratio adjustment on progressive input. Alternatively, get a player with an *aspect ratio control* option that "windowboxes" 4:3 video into a 16:9 rectangle by squeezing it horizontally and adding black bars on the side. Because of the added scaling step, this may reduce picture quality, but at least it gets around the problem.

Just as early DVD computers did a poor job of displaying progressive-scan DVDs, the first generations of progressive consumer players are also a bit disappointing. But as techniques improve, as DVD producers become more aware of the steps they must take to ensure a good progressive display, and as more progressive displays appear in homes, the experience will undoubtedly improve, making home theaters more like real theaters.

For more on progressive video and DVDs, see Part 5, "Progressive Scan DVD," at "Secrets of Home Theater and High Fidelity" (www.hometheaterhifi.com/volume_7_4/dvd-benchmark-part-5-progressive-10-2000.html).

Why Doesn't Disc X Work in Player Y?

The DVD specification is complex and open to interpretation. DVD-Video title authoring is also very complex. As with any new technology, there are compatibility problems. The DVD-Video standard has not changed substantially since it was finalized in 1996, but many players don't properly support it. Discs have become more complex as authoring tools improve, so recent discs often uncover engineering flaws in players. Some discs behave strangely or won't play at all in certain players. In some cases, manufacturers can fix the problem with an upgrade to the player (see "What's Firmware and Why Would I Need to Upgrade It?"). In other cases, disc producers need to reauthor the title to correct an authoring problem or to work around a player defect. Problems can also occur because of damaged or defective discs or because of a defective player.

If you have problems playing a disc, try the following:

1. Check the following Table 1-1 to see if it's a reported problem. Also check the list of problem discs in DVD Review's Film Vault and at InterActual's tech support page. Or try a newsgroup search at Google.

2. Try playing the disc a few more times. If you don't get the exact same problem every time, it's probably a defective or damaged disc. Make

sure the disc isn't dirty or scratched (see "How Should I Clean and Care for DVDs?").

3. Try the disc in a different player. (Visit a friend or a nearby store that sells players.) If the disc plays properly in a different player, your player is likely at fault. Contact the manufacturer of your player for a firmware upgrade. Or, if you bought the player recently, you may want to return it for a different model.

4. Try a different copy of the disc. If the problem doesn't recur, it indicates that your first copy was probably damaged or defective. If more than one copy of the disc has problems in more than one player, it may be a misauthored disc. Contact the distributor or the studio about getting a corrected disc.

For other DVD and home theater problems, try Doc DVD or DVD Digest's Tech Support Zone. If you have a Samsung 709, see the Samsung 709 FAQ. For troubleshooting DVDs on computers, see Chapter 4's "Why Do I Have Problems Playing DVDs on My Computer?" The Dell Inspiron 7000 DVD Movie List (www.eaglecomputing.com/dvdlist.htm) has Inspiron-specific problems.

Table 1-1 outlines the problems reported by readers of the DVD FAQ. The author of this book has not verified these claims and takes no responsibility for their accuracy.

How Do the Parental Control and Multirating Features Work?

DVDs include parental management features for blocking playback and for providing multiple versions of a movie on a single disc. Players (including software players on PCs) can be set to a specific parental level using the onscreen settings. If a disc with a rating above that level is put in the player, it won't play. In some cases, different programs on the disc have different ratings. The level setting can be protected with a password.

A disc can also be designed so that it plays a different version of the movie depending on the parental level set in the player. By taking advantage of the seamless branching features of DVD, objectionable scenes are automatically skipped over or replaced during playback. This requires that the disc be carefully authored with alternate scenes and branch points that don't cause interruptions or discontinuities in the soundtrack. No standard way exists for identifying which discs have multirated content.

TABLE 1-1 DVD Problems

Title	Player	Problem	Solution
Various Polygram Titles	Early Toshiba and Magnavox models	Won't load or freezes	Upgrade available from Toshiba service centers
Various Central Park Media (anime) titles	Similar problems as *The Matrix*		
Any all-region title	Many JVC models	Rejects disc	Unknown
RCE titles	Fisher DVDS-1000, Sanyo Model DVD5100	World map and "only plays on nonmodified players" message	Contact Sanyo/Fisher tech support for workaround
The Abyss, Special Edition	Early Toshiba models	Disc 2 won't load or freezes	Upgrade available from Toshiba service centers
	Many cheap players	Repeats scenes	Player doesn't properly handle seamless branching so get upgrade from manufacturer
	Apex AD-600A	Scenes play twice	Check with Apex for upgrade
AI (PAL region 2)	Wharfdale 750	Won't play	Unknown
Akira, Special Edition	Pioneer DV-37, DV-737, DV-525	Freezes in several places	Fast forward to skip trouble spots
Aliens 20th Anniversary Edition	Pioneer DV-S737	Picture degrades after layer change	Unknown
American Beauty Awards Edition	Toshiba SD-3108, Philips DVD805	Won't load	Upgrade from manufacturer service center (Toshiba firmware 3.30 or newer)
American Pie	Philips 940	Freezes at layer change (1:17:09)	Unknown
Any Given Sunday	Pioneer Elite DVL90	Won't load	Upgrade from Pioneer service center
Arlington Road	See *Cruel Intentions*		
Armageddon	Panasonic A115-U and A120-U	Won't load	Unplug player with disc inserted, then plug it in, and turn it on

(continued)

TABLE 1-1   DVD Problems (continued)

Title	Player	Problem	Solution
Avengers TV series (A&E)	Toshiba SD-3108	Locks up player	Upgrade available from Toshiba service centers
	Philips 930, 935	Won't load	Check with Philips for firmware upgrade
Back to the Future Trilogy (region 4)	Various players	"Anecdote" subpictures don't play properly	Unknown
Bats	Apex AD 600A	Won't load	Check with Apex for upgrade
Big Trouble in Little China Special Edition	Panasonic SC-DK3	Won't load	Unplug player with disc inserted, then plug it in, and turn it on
The Blair Witch Project	Some Toshiba players	Doesn't play properly	Upgrade available from Toshiba service centers
Cruel Intentions	Some JVC and Yamaha players	An error in first release messes up the parental controls, causing other discs to not play	Reset the player or get the corrected version of the disc or set the parental country code to AD with a password of 8888
Deep Blue Sea	Similar problems as The Matrix		
Dinosaur	Many players (JVC-XV501BK, Philips DVD781 CH, Pioneer DV-737/ DV-37/DV-09/ DVL-919/DV-525/ DVL-90/KV-301C, Sony 7700, Panasonic A300, Toshiba SD-3109, RCA 5220, Denon DVD 2500, Magnavox DVD502AT, Toshiba 2109/3109, JVC XV-D2000/ V-D701, Oritron DVD600/DVD100, Sylvania DVL100A, and others)	Won't load, ejects disc, freezes, skips, slow menus, won't pause/forward/rewind, or sound cuts out	Authoring problem, so contact Disney for a replacement (also see Disney's The Kid below)
The Kid	Many players (Apex 600AD, Philips 711, Pioneer DV-737, RCA, and others)	Skips, ejects disc, freezes, blue lines onscreen	Authoring problem, so contact Disney for a replacement (solution on Philips player: put disc in player drawer

Title	Player	Problem	Solution
The Kid (continued)			but do not close drawer, and press "1" on remote to jump to Chapter 1)
Dragon's Lair	Toshiba SD2109/3109 (before mid-1999)	Various	Upgrade available from Toshiba service centers
	Most Samsung, Aiwa	Various	Check with Samsung (800-726-7864) or Aiwa for firmware upgrade
Enigma, 2002	Toshiba SD-4700	Won't play	Unknown
Entrapment	JVC, Sony 850	Freezes	Check with JVC for firmware upgrade
	Sigma Hollywood Plus	See *The World Is Not Enough*	
Everything, Everything (Underworld)	Toshiba SD3108 and SD3109	Won't load	Upgrade available from Toshiba service centers
Evolution	Many computer DVD software players	Won't play	Contact studio for new version of disc
Galaxy Quest	Most Samsung players	Freezes at Chapter 7	Check with Samsung (800-726-7864) for firmware upgrade
Girl, Interrupted remote available	Apex AD-600A, Shinco 2120, Smart DVDMP3000, others	Jumps to Features menu, won't play movie	Press Resume on control; upgrade for Smart
Gladiator	Toshiba SD3108/SD3109, Wharfedale DVD 750, others	Won't load	Contact studio for new version of disc
The Godfather Collection bonus disc	A few players	Various problems	Upgrade your player or get new disc from Paramount (replacement disc works around player bugs)
Good Will Hunting	Apex AD-3201	Won't play audio commentary	
Idle Hands	See *Cruel Intentions*		

(continued)

TABLE 1-1   DVD Problems (continued)

Title	Player	Problem	Solution
In the Heat of the Night	Pioneer Elite DVL-90	Won't play	
Independence Day	Toshiba SD3108 and SD3109	Won't load	Upgrade available from Toshiba service centers
	Philips DVD805 and DVD855	Won't load	Check for upgrade from Philips
	Many cheap players	Repeats scenes	Player doesn't properly handle seamless branching, so get upgrade from manufacturer
Insomnia	Toshiba SD1700	Stutters and freezes	Unknown
The Last Broadcast	GE 1105P	Won't load	Unknown
The Last of the Mohicans	See *The World Is Not Enough*		
Lord Peter Wimsey: The Nine Taylors	Yamaha DVD-C900	Disc 2 won't load or freezes in menu	
Lost In Space	Sharp	Freezes	
	Creative DXR3	Freezes, audio out of sync	Check for updated drivers
The Man With The Golden Gun	A few first-generation players, many software player	Garbled video after layer change	Might be a disc authoring error
The Matrix	Various players	Various problems	Details at InterActual tech support (for GE 1105-P, serial number beginning with 940 or lower, so get upgrade from GE; see Samsung 709 FAQ)
Mission Impossible II	Toshiba SD-3108	Won't load	Get upgrade from manufacturer service center
Mission to Mars	Toshiba SD-3108	Won't load	Get upgrade from manufacturer service center

Title	Player	Problem	Solution
Monsters, Inc.	Various players	Locks up near end of movie	Seems to be player flaws, so check for player upgrade; Disney may reauthor disc with a workaround
The Mummy	Philips 930, 935	Won't load	Unknown
The Mummy Returns	Zenith DVD 2200	Video skewed left or right on bonus material	Unknown
The Patriot	Apex AD 600A	Won't play movie	Check with Apex for upgrade (pressing "resume" may work)
	JVC XV-511BK	Won't load	Check with JVC for upgrade
The Perfect Storm	Toshiba SD-3108	Won't load	Get upgrade from manufacturer service center
Planet of the Apes	Toshiba SD-2109	PIP feature activates and locks up when the two ape generals fight	Unknown
The Princess Bride Special Edition	Toshiba SD-3109	Freezes during first sword fight scene	Unknown
Saving Private Ryan	All players	Distortion (smearing, flares) in beach scene at end of Chapter 4	This is a deliberate camera effect in the film
Scary Movie	Creative Encore 12x, GE 1105P	Crashes in FBI warning	Try to skip past FBI warning or check for bug fix from Creative
The Simpsons, The Complete Second Season	Yamaha DVD-C900	Some special features on disc 4 cause player to crash	Unknown
The Sixth Sense update	Sigma Hollywood Plus	MMSYSTEM275 error	Wait for a software from Sigma
Sleepy Hollow	Some Toshiba players	Doesn't play properly	Upgrade available from Toshiba service centers

(continued)

TABLE 1-1   DVD Problems (continued)

Title	Player	Problem	Solution
Snow White	Windows 2000 and Windows XP	Doesn't play movie	Fix available from Microsoft
Space Ace	See Dragon's Lair		
Stargate Special Edition	Magnavox 400AT	Freezes in director's commentary	Unknown
Stuart Little	See Girl Interrupted		
Three Kings	LG DVD-2310P	Won't play extras	Unknown
Thomas the Tank Engine	See Girl Interrupted		
Tomorrow Never Dies	Sharp 600U Bush DVD2000	Locks up player, won't load	Unknown
Universal Soldier	Wharfedale 750	Picture breakup after Chapter 30	Might be a problem with the disc
Wild Wild West	Samsung DVD 709; Philips 930, 935; GE 1105P	Won't load	Check with Samsung (800-726-7864), Philips, or GE for firmware upgrade
The World Is Not Enough	Sigma Hollywood Plus	MMSYSTEM275 error	Wait for a software update from Sigma, might be related to trying to play in wrong region
The World Is Not Enough (region 2)	Philips 750	Stutters and freezes	Presumably a flaw in the player but plays region 1 version okay
You've Got Mail	Various players	Various problems	Details at InterActual tech support

Unfortunately, very few multirating discs have been produced. Hollywood studios are not convinced that a big enough demand exists to justify the extra work involved (shooting extra footage, recording extra audio, editing new sequences, creating branch points, synchronizing the soundtrack across jumps, submitting new versions for MPAA ratings, dealing with players that don't properly implement parental branching, having video store chains refuse to carry discs with unrated content, and much more). If this feature is important to you, let the studios know. A list of studio addresses is available at DVD File, and a Studio and Manufacturer Feedback area is provided at Home Theater Forum. You might also want to visit the Viewer Freedom site.

Multiratings discs include *Kalifornia, Crash, Damage, Embrace of the Vampire, Poison Ivy,* and *Species II.* In most cases, these discs provide "uncut" or unrated versions that are more intense than the original theatrical release. Discs that use multistory branching (not always seamless) for a director's cut or special edition version include *Dark Star, Stargate SE, The Abyss, Independence Day,* and *Terminator 2 SE* (2000 release). Also see www.multipathmovies.com.

Another option is to use a software player on a computer that can read a playlist telling it where to skip scenes or mute the audio. Playlists can be created for the thousands of DVD movies that have been produced without parental control features. ClearPlay seems to be the most successful product of this type. A shareware Cine-bit DVD Player did this, but it has been withdrawn apparently because of legal threats from Nissim, who seem determined to stifle the very market they claim to support. A Canadian company, Select Viewing, is releasing software for customized DVD playback on Windows PCs. A few similar projects are under development.

Yet another option is *TV*Guardian or Curse Free TV, a device that is attached between the DVD player and the TV to filter out profanity and vulgar language. The box reads the closed caption text and automatically mutes the audio and provides substitute captions for objectionable words. (Note that current versions of these devices don't work with digital audio connections, and don't work with DVDs without NTSC Closed Captioning.)

Which Discs Include Multiple Camera Angles?

A euphemism is used in the DVD industry, where "multiangle titles" — spoken with the right inflection — means adult titles. However, apart from thousands of XXX-rated discs, not very many mainstream DVDs have multiple angles, because it takes extra work and limits playing time (a segment with two angles uses up twice as much space on the disc).

Short Cinema Journal Volume 1 was one of the first to use camera angles in the animated "Big Story," which is also available on the *DVD Demystified* first edition sample disc. *Ultimate DVD* (Gold or Platinum) is another sample disc with examples of angles. *King Crimson: Deja Vroom* has excellent angles, allowing you to focus on any of the musicians. Other multiangle music discs include *Dave Matthews Band: Listener Supported,*

Metallica: Cunning Stunts, and *Sarah McLachlan: Mirrorball.* Some movies, such as *Detroit Rock City, Ghostbusters Special Edition, Mallrats, Suicide Kings, Terminator 2 Special Edition,* and *Tomorrow Never Dies Special Edition* use multiple angles in supplements. Some discs, especially those from Buena Vista, use the angle feature to show credits in the selected language (usually with the angle button locked out).

You can get an incomplete list of multiangle discs by doing an extended search at DVD File or a power search at DVD Express. To weed out the adult titles at DVD Express, select all entries in the category list (click top entry and Shift-click bottom entry) and then deselect Adult (Ctrl-click).

Is It Okay to Put Labels or Magnetic Strips on DVDs?

Labels and adhesive strips are dangerous because they can unbalance the disc and cause errors, or even damage a player, especially if they peel off while the disc is spinning. Pressure-sensitive adhesives break down over time, so it's possible for labels to come loose after a few years. Libraries and DVD rental outlets often want to label discs or attach magnetic strips for security, but it's best not to use them at all. If you must, use a ring-shaped "donut" label that goes around the center of the disc. As long as the circular label doesn't interfere with the player clamping onto the hub, it should be okay. If you have to use a noncircular sticker, place it as close to the center as possible to minimize any unbalancing. Placing a second sticker straight across from the center will also help.

Writing with a marker in the clear (not reflective) area at the hub is better than using a sticker, although not much room to write is provided. Write only in the area inside a 44-millimeter diameter. Writing anywhere else on the disc is risky, because the ink could possibly eat away the protective coating and damage the data layer underneath.

In most cases, a better alternative for security is a case that can only be opened with special equipment at the register or checkout counter. Barcodes, stickers, and security strips can be placed on the case without endangering discs (or players). This is especially good for double-sided discs, which have no space for stickers.

Full-size round labels designed to go on recordable CDs and DVDs can be used, but they have been known to cause problems. As DVD-ROM drives get faster and faster, destabilization of the disc by a label may cause read errors. A better (but more expensive) solution is to use an inkjet disc printer (IMT, Odixion, Primera, Rimage, or Trace Affex) with printable-surface discs.

What's the Difference Between Closed Captions and Subtitles?

Closed Captions are a standardized method of encoding text into an NTSC television signal. The text can be displayed by a TV with a built-in decoder or by a separate decoder. All TVs larger than 13 inches sold in the United States since 1993 have Closed Caption decoders. Closed Captions can be carried on DVD, videotape, broadcast TV, cable TV, and so on.

Even though the terms *caption* and *subtitle* have similar definitions, *captions* commonly refer to onscreen text specifically designed for hearing-impaired viewers, while *subtitles* are straight transcriptions or translations of the dialogue. Captions are usually positioned below the person who is speaking, and they include descriptions of sounds (such as gunshots or closing doors) and music. *Closed* captions are not visible until the viewer activates them. *Open* captions are always visible, such as subtitles on foreign videotapes.

Closed Captions on DVDs are carried in a special data channel of the MPEG-2 video stream and are automatically sent to the TV. You can't turn them on or off from the DVD player. Subtitles, on the other hand, are DVD subpictures, which are full-screen graphical overlays (see "What Are the Video Details?" in Chapter 3 for technical details). One of up to 32 subpicture tracks can be turned on to show text or graphics on top of the video. Subpictures can also be used to create captions. To differentiate between NTSC Closed Captions and subtitles, captions created as subpictures are usually called "captions for the hearing impaired."

If this is all too confusing, just follow this advice: To see Closed Captions, use the CC button on the TV remote. To see subtitles or captions for the hearing impaired, use the subtitle button on the DVD remote or use the onscreen menu provided by the disc. Don't turn both on at once or they'll end up on top of each other. Keep in mind that not all DVDs have Closed Captions or subtitles. Also, some DVD players don't reproduce Closed Captions at all.

See DVD File's "A Guide to DVD Subtitles and Captioning," Gary Robson's Caption FAQ, and Joe Clark's DVD Accessibility for more about Closed Captions. Note that DVD does not support PAL Teletext, the much-improved European equivalent of Closed Captions.

What Do the "D" Codes on Region 2 DVDs Mean?

Some non-U.S. discs from Warner, MGM, and Disney are marked with a distribution zone number. D1 identifies a UK-only release. These often have

English-only soundtracks with BBFC censoring. D2 and D3 identify European DVDs that are not sold in the United Kingdom and Ireland. These often contain uncut or less cut versions of films. D4 identifies DVDs that are distributed throughout Europe (region 2) and Australia/New Zealand (region 4).

What's Firmware and Why Would I Need to Upgrade It?

DVD players are simple computers. Each one has a software program that controls how it plays discs. Because the software is stored on a chip, it's called firmware. Some players have flaws in their programming that cause problems playing certain DVDs. In order to correct the flaws, or in some cases to work around authoring errors on popular discs, the player must be upgraded with a replacement firmware chip. This usually has to be done in a factory service center, although some players can be upgraded simply by inserting a CD. Refer to "Why Doesn't Disc X Work in Player Y?" for more on compatibility problems.

Can Discs Help Me Test, Optimize, or Show Off My Audio/Video System?

A few DVDs are designed specifically for testing and optimizing video and audio playback, and some demonstrate special DVD features:

- *AVIA Guide to Home Theater* by Ovation Software (extensive video and audio test patterns and setup tutorials, www.ovationsw.com)
- *Video Essentials* by Joe Kane Productions (the original system optimization disc, from the master, www.videoessentials.com)
- *Ultimate DVD Series* by Henninger Interactive (examples of many DVD features, plus test and demo material, www.henniger.com)
- *DVD Demystified* demo disc (examples of almost every DVD feature, plus demo material, dvddemystified.com/disc.html)

Here are a few movies that work especially well for demonstrating DVD's video and audio quality:

- ***Dinosaur***   Direct-to-DVD digital transfer gives sharp, clear images, with good bass on footsteps and fights.
- ***The Eagles: Hell Freezes Over***   Outstanding 5.1-channel music (DTS only and Dolby Digital tracks are 2-channel).
- ***The Fifth Element***   Excellent video, especially in beginning desert scenes, with stellar audio as well.

- *Gladiator* Stunning surround audio with brilliantly mixed orchestration.

- *O Brother, Where Art Thou* Beautiful color and incredible detail (check out facial stubble) with well-rendered shadows.

- *Terminator 2: Judgment Day (Ultimate Edition)* Great video for shadows and reds, with highly dimensional audio.

- *Toy Story 2* The perfect all-digital transfer results in sharp, rich images; the sound effects are nicely staged.

- *U-571* Its earthshaking bass makes it a great subwoofer demo.

Films on Disc has a list of ISF DVD *citations*, examples of the best of the craft.

What Do Sensormatic and Checkpoint Mean?

Sensormatic and Checkpoint are two point-of-sale security systems. The names refer to the little metal tags inserted into DVD packaging to set off an alarm if you go through the sensors at the store entrance without having the tags deactivated during checkout. The tags are placed in the packages at the replication plant so that it doesn't have to be done at the store. This is called *source tagging*.

What Are Superbit, Infinifilm, and Other Variations of DVD?

One single DVD-Video standard exists, but within the DVD-Video format is a great deal of flexibility in the way discs can be created. Different studios have come up with brand names for their particular implementations of advanced features. Nothing is extraordinary about any particular variation, other than a studio spending a lot of time and effort making it work well and promoting it. These kinds of advanced DVDs should play on most players but may reveal more player bugs than standard discs (see "Why Doesn't Disc X Work in Player Y?").

Superbit DVDs, from Columbia TriStar, use a high data rate for the video to improve picture quality. Additional language tracks and other extras are left off the disc to make room for more video data and for a DTS audio track. In most cases, the difference is subtle, but it does improve the experience on high-end players and progressive-scan displays. See superbitdvd.com for marketing info.

Infinifilm DVDs, from New Line, let you watch a movie with pop-ups that direct you to extra content such as interviews, behind-the-scenes footage, or historical information. See infinifilm.com for more hype.

I Don't Know the Parental Control Password for My Player. What Do I Do?

Most DVD players enable you to lock out discs above a certain rating (refer to the parental rating section). The rating level is protected by a password so that children (or spouses) can't change it. If you don't know the password, you won't be able to play some discs. You might be able to clear the password by resetting the player (see the user manual) or by unplugging it for a few days. In some cases, you might be able to use the default password (0000, 9999, or 3308). Otherwise, you'll have to call the customer service number of the manufacturer and see if they can help you.

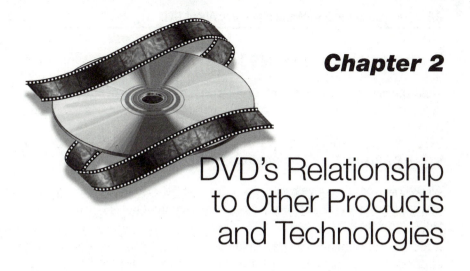

DVD's Relationship to Other Products and Technologies

Will DVD Replace VCR?

Eventually. DVD player sales exceeded VCR sales in 2001. DVD recorders are also available (refer to "Can DVDs Record from VCRs, TVs, and So On?" in Chapter 1, "General DVD") and will hasten the death of VCRs once the price difference is small enough. DVDs have many advantages over VHS tapes, such as no rewinding, quick access to any part of a recording, and fundamentally lower technology costs for hardware and disc production. Some projections show DVD recorder sales passing VCR sales in 2005. By 2010, VHS may be as dead as vinyl records were in 2000.

Will DVD Replace CD-ROM?

Most CD-ROM drive manufacturers plan to cease CD-ROM drive production at some point, in favor of DVD-ROM drives. Because DVD-ROM drives can read CD-ROMs, a compatible forward migration path exists. CD-ROM discs will continue to be used in cases where the extra capacity isn't needed, because they cost less to make than DVD-ROM discs.

Can CD-R Writers Create DVDs?

No, because DVDs use a laser with a smaller wavelength to allow smaller pits in tracks that are closer together. The DVD laser must also focus more tightly and at a different level. In fact, a disc made on a current *CD-recordable* (CD-R) writer may not be readable by a DVD-ROM drive (see "Is CD-R Comatible with DVD"). It's unlikely that CD-R drives will be upgraded to *DVD-Recordable* (DVD-R) drives, because this would cost more than purchasing a new DVD-R drive.

Is CD Compatible with DVD?

This actually has many answers, covered in the following sections. Note the differentiation between DVDs (general media) and DVD-ROMs (computer data).

Is CD Audio (CD-DA) Compatible with DVDs?

All DVD players and drives will read audio CDs (Red Book). This is not actually required by the DVD spec, but so far all manufacturers have made their DVD hardware able to read CDs.

On the other hand, you can't play a DVD in a CD player. The pits are smaller, the tracks are closer together, the data layer is a different distance from the surface, the modulation is different, and the error correction coding is new. Also, you can't put CD audio data onto a DVD and have it play in DVD players (Red Book audio frames are different from DVD data sectors.)

Is CD-ROM Compatible with DVD-ROM?

All DVD-ROM drives will read CD-ROMs (Yellow Book), and software on a CD-ROM will run fine in a DVD-ROM system. However, DVD-ROMs are not readable by CD-ROM drives.

Is CD-R Compatible with DVD?

Sometimes. The problem is that most CD-R discs (Orange Book Part II) are "invisible" to the DVD laser wavelength because the dye used to make the blank CD-R doesn't reflect the beam. Some first-generation DVD-ROM drives and many DVD players can't read CD-R discs. The formulation of dye used by different CD-R manufacturers also affects readability. That is, some brands of CD-R discs have better reflectivity at the DVD laser wavelength, but even these don't reliably work in all players. An effort to develop CD-R Type II media compatible with both CD and DVD wavelengths has been abandoned.

The common solution is for the DVD player or drive to use two lasers at different wavelengths: one for reading DVDs and the other for reading CDs and CD-Rs. Variations on the theme include Sony's dual discrete optical pickup with switchable pickup assemblies that have separate optics, dual-wavelength lasers (initially deployed on Sony's PlayStation 2), and Samsung's annular masked objective lens with a shared optical path. Other approaches include Toshiba's shared optical path using an objective lens masked with a coating that's transparent only to 650-nanometer light,

Hitachi's switchable objective lens assembly, and Matsushita's holographic dual-focus lens. The MultiRead logo guarantees compatibility with CD-R and *CD-rewritable* (CD-RW) media, but unfortunately few manufacturers are using it. The bottom line is that if you want a DVD player that can read CD-R discs, look for a dual laser, twin laser, or dual optics feature.

DVD-ROM drives can't record on CD-R discs or any other media, but a few combination DVD-ROM/CD-RW drives can write to CD-R and CD-RW discs. Most newer recordable DVD drives (see "What About Recordable DVDs: DVD-R, DVD-RAM, DVD-RW, DVD+RW, and DVD+R?" in Chapter 4, "DVDs and Computers") can also record on CD-R or CD-RW discs. CD-R burners, however, can't read or write DVD discs of any kind.

Is CD-RW Compatible with DVD?

Usually. CD-RW discs (Orange Book Part III) have a smaller reflectivity difference, requiring new *automatic gain control* (AGC) circuitry in CD-ROM drives and CD players. Most existing CD-ROM drives and CD players can't read CD-RW discs. The multiread standard addresses this, and some DVD manufacturers have suggested they will support it. The optical circuitry in even first-generation DVD-ROM drives and DVD players is usually able to read CD-RW discs, because CD-RW discs do not have the "invisibility" problem of CD-R discs (refer to the previous section). Most newer recordable DVD drives (see Chapter 4) can also record on CD-R or CD-RW. CD-RW burners can't read or write DVD discs of any kind.

Are Video CDs Compatible with DVD Players?

Sometimes. It's not required by the DVD spec, but it's trivial to support the Video CD (White Book) standard because any MPEG-2 decoder can also decode MPEG-1 from a Video CD. About two-thirds of DVD players can play Video CDs, including most Panasonic, RCA, Samsung, and Sony models. Japanese Pioneer models play Video CDs, but American models older than the DVL-909 don't. Also, Toshiba players older than models 2100, 3107, and 3108 don't play Video CDs.

Video CD resolution is 352×288 for PAL and 352×240 for NTSC. The way most DVD players and Video CD players deal with the difference is to chop off the extra lines or add blank lines. When playing PAL Video CDs, the Panasonic and RCA NTSC players apparently cut 48 lines (17 percent) off the bottom, whereas Sony NTSC players scale all 288 lines to fit.

Because PAL Video CDs are encoded to play 24 *frames per second* (fps) film at 25 fps, usually a 4 percent speedup occurs. The playing time is shorter, and the audio is shifted up in pitch unless it was digitally processed before encoding to shift the pitch back to normal. This also happens with PAL DVDs (see "Is DVD-Video a Worldwide Standard? Does It Work with NTSC, PAL, and SECAM?" in Chapter 1).

All DVD-ROM computers can play Video CDs (with the right software). Standard Video CD players can't play DVDs.

NOTE: Many Asian Video CDs achieve two soundtracks by putting one language on the left channel and another on the right. The two channels are mixed together into overlapping sound on a stereo system unless you adjust the balance or disconnect one input to get only one channel.

For more on Video CD, see Glenn Sanderse's Video CD FAQ at CDPage, or Russil Wvong's Video CD FAQ.

Are Super Video CDs Compatible with DVD Players?

Not generally. *Super Video CD* (SVCD) is an enhancement to Video CD that was developed by a Chinese government-backed committee of manufacturers and researchers, partly to sidestep DVD technology royalties and partly to create pressure for low DVD player and disc prices in China. The final SVCD spec was announced in September of 1998, winning out over C-Cube's *China Video CD* (CVD) and HQ-VCD (from the developers of the original Video CD).

In terms of video and audio quality, SVCD falls between Video CD and DVD, using a 2x CD drive to support 2.2 Mbps VBR MPEG-2 video (at 480×480 NTSC or 480×576 PAL resolution) and two-channel MPEG-2 Layer II audio. As with DVD, it can overlay graphics for subtitles. It's technically easy to make a DVD-Video player compatible with SVCD, but it's being done mostly on Asian DVD player models. The Philips DVD170 player can be upgraded (using a special disc) to play SVCD discs. SVCD players can't play DVDs, however, because the players are based on CD drives. See Jukka Aho's Super Video CD Overview (www.iki.fi/znark/video/svcd/overview) and Super Video CD FAQ (www.iki.fi/znark/video/svcd/faq) for more info.

Are Picture CDs or Photo CDs Compatible with DVD Players?

Sometimes. Because Picture CDs and Photo CDs are usually on CD-R media, they suffer from the CD-R problem (refer to "Is CD-R Compatible with DVD?"). That aside, some DVD players can play Picture CDs. Only a few can play Photo CDs.

Most DVD-ROM drives will read Picture CDs or Photo CDs (if they read CD-R discs) because it's trivial to support the XA and Orange Book multi-session standards. Picture CDs are designed to work with Windows, whereas Photo CDs require specific support from an application or an OS. Photos can be put on recordable DVDs using the DVD-Video slideshow feature, which works on all DVD players (see "How Do I Copy My Home Videos/Movies/Slides to DVDs?" in Chapter 5, "DVD Production".)

Is CD-i Compatible with DVD?

In general, no. DVD players do not play CD-i (Green Book) discs. Philips once announced that it would make a DVD player that supported CD-i, but it never appeared. Some people expected Philips to create a DVD-i format in an attempt to breathe a little more life into CD-i (and recover a bit more of the billion dollars or so they invested in it). A DVD-ROM PC with a CD-i card should be able to play CD-i discs.

CD-i movies use the CD-i digital video format that was the precursor to Video CD. Early CD-i DV discs won't play on DVD players or VCD players, but newer CD-i movies, which use the standard VCD format, will play on any player that can play VCDs (refer to "Are Video CDs Compatible with DVD Players?"). See Jorg Kennis' CD-i FAQ for more information on CD-I (www.icdia.org/faq).

Are Enhanced CDs Compatible with DVD Players?

DVD players will play music from enhanced music CDs (Blue Book, CD Plus, or CD Extra), and DVD-ROM drives will play music and read data from enhanced CDs. Older enhanced CD formats such as mixed mode and track zero (pregap or hidden track) should also be compatible, but a problem occurs with Microsoft and other CD/DVD-ROM drivers skipping track zero.

Is CD+G Compatible with DVD Players?

Only a few players, such as the Pioneer DVL-9 and Pioneer karaoke DVD models DV-K800 and DVK-1000, support CD+G. Most DVD players don't support this mostly obsolete format. All DVD-ROM drives can read the CD+G information, but special software is required to make use of it.

Is CDV Compatible with DVD?

Sort of. CDV, sometimes called video single, is actually a weird combination of a CD and a laserdisc. Part of the disc contains 20 minutes of digital audio playable on any CD or DVD player. The other part contains five minutes of analog video and digital audio in laserdisc format, playable only on a CDV-compatible laserdisc player. Pioneer's combination DVD/laserdisc players are the only DVD players that can play CDVs. Of course, standard laserdisc/CDV players can't play DVDs (see "Are Laserdiscs Compatible with DVD Players?" for more info.)

Is MP3 Compatible with DVD?

Not officially. MP3 is the MPEG Layer 3 audio compression format. (MP3 is not MPEG-3, which doesn't exist.) The DVD-Video spec allows only Layer 2 for MPEG audio (MP2). However, MP3 can be played by any computer with a DVD-ROM drive, and many DVD players (particularly those manufactured in Asia) can play MP3 CDs. However, oddly enough, most of the players that can play MP3s from a CD can't play MP3s from a DVD.

Is HDCD Compatible with DVDs?

Yes. Pacific Microsonics' (www.hdcd.com) *high-definition compatible digital* (HDCD) format is an encoding process that enhances audio CDs so that they play normally in standard CD and DVD players (and allegedly sound better than normal CDs). Yet they produce an extra 4 bits of precision (20 bits instead of 16) when played on CD and DVD players equipped with HDCD decoders.

Are Laserdiscs Compatible with DVD Players?

Standard DVD players will not play laserdiscs, and you can't play a DVD disc on any standard laserdisc player. (Laserdiscs use analog video, whereas DVDs use digital video; they are very different formats.) Pioneer makes combo players that play laserdiscs and DVDs (as well as CDVs and audio CDs).

Will DVDs Replace Laserdiscs?

When this question was first asked in 1996, before DVD was even available, people wondered if DVDs would replace laserdiscs. Some argued it never would, that DVD would fail and its adherents would come groveling back to laserdiscs. After DVDs were released, it soon became clear that they had doomed laserdiscs to quick obscurity. Pioneer Entertainment, the long-time champion of laserdiscs, abandoned them in June of 1999. This was sooner than even Pioneer thought possible (in September 1998, Pioneer's president Kaneo Ito said the company expected laserdisc products to be in the market for another one-and-a-half to two years).

Laserdiscs fill niches in education and training, but they are fading even there. Existing players and discs will be around for a while, though almost no new discs are being produced. Once over 9,000 laserdisc titles were in the U.S. market and a total of over 35,000 titles worldwide could be played on over 7 million laserdisc players. It took DVDs several years to reach this level, and certain rare titles are available on laserdisc but not on DVD. One bright point is that laserdiscs can now be had at bargain prices.

How Do DVDs Compare to Laserdiscs?

To answer this question, let's examine some different criteria:

- **Features**   DVDs have the same basic features as CLV laserdiscs (scan, pause, and search) and CAV LD (freeze and slow), and they add branching, multiple camera angles, parental control, video menus, and interactivity, although some of these features are not available on all discs.

- **Capacity**   Single-layer DVDs hold over two hours; dual-layer DVDs holds over four hours. CLV laserdiscs hold one hour per side, CAV laserdiscs hold half an hour, along with 104,000 still images. A DVD can hold thousands of still pictures, accompanied by hundreds of hours of audio and text.

- **Convenience**   An entire movie fits on one side of a DVD, so you don't have to flip the disc or wait for the player to do it. DVDs are smaller and easier to handle, and DVD players can be portable, similar to CD players. Discs can be easily and cheaply sent through the mail. On the other hand, laserdiscs have larger covers for better art and text.

- **Noise**   Most laserdisc players make a whirring noise that can be heard during quiet segments of a movie. Most DVD players are as quiet as CD players.

- **Audio**   Laserdiscs can have better quality with Dolby Surround soundtracks in an uncompressed PCM format. DVDs have better quality of Dolby Digital or music only (PCM). Laserdiscs have two audio tracks, analog and digital, whereas DVDs have up to eight audio tracks. Laserdiscs use PCM audio sampled with 16 bits at 44.1 kHz. DVD LPCM audio can use 16-, 20-, or 24-bit samples at 48 or 96 kHz (although PCM is not used with most movies). Laserdiscs have surround audio in Dolby Surround, Dolby Digital (AC-3), and *Digital Theater Systems* (DTS) formats; 5.1-channel surround sound is available by using one channel of the analog track for AC-3 or both channels of the digital track for DTS. DVDs use the same Dolby Digital surround sound, usually at a higher data rate of 448 kbps, and can optionally include DTS (at data rates up to 1536 kbps compared to laserdiscs' 1411 kbps, but in practice DTS data rates are often 768 kbps). DVD players convert Dolby Digital to Dolby Surround. The downmixing, combined with the effects of compression, often results in lower-quality sound than from laserdisc Dolby Surround tracks.

- **Video** DVDs usually have better video. Laserdiscs suffer from degradation inherent in analog storage and in the composite NTSC or PAL video signal. DVDs use digital video, and even though it's heavily compressed most professionals agree that, when properly and carefully encoded, it's virtually indistinguishable from studio masters. This doesn't mean that the video quality of a DVD is always better than a laserdisc. Only that it can be better. Also keep in mind that the average television is of insufficient quality to show much difference between laserdiscs and DVDs. Home theater systems or *high-definition TVs* (HDTVs) are needed to take full advantage of the improved quality.

- **Resolution** In numerical terms, a DVD has 345,600 pixels (720×480), which is 1.3 times the 272,160 pixels of a laserdisc (567×480). A widescreen DVD has 1.7 times the pixels of a letterboxed laserdisc (or 1.3 times an anamorphic laserdisc). As for lines of horizontal resolution, a DVD has about 500, whereas a laserdisc has about 425 (more info is covered in "What Do Lines of Resolution Mean?" in Chapter 3, "DVD Technical Details"). In analog output signal terms, a typical luma frequency response maintains a full amplitude between 5.0 and 5.5 MHz. This is below the 6.75 MHz native frequency of the MPEG-2 digital signal. The chroma frequency response is half that of luma, and the laserdisc frequency response usually begins to fall off at 3 MHz. (All figures are for NTSC, not PAL.)

- **Legacy titles** Some movies on laserdisc will probably never appear on DVD.

- **Availability** DVD players and discs are available for purchase and rental in thousands of outlets and on the Internet. Laserdisc players and discs are becoming hard to find.

- **Price** Low-cost DVD players are cheaper than the cheapest laserdisc player. Most movies on DVD cost less than on laserdisc.

- **Restrictions** For those outside the United States, regional coding (refer to "What Are Regional Codes, Country Codes, or Zone Locks?" in Chapter 1) is a definite drawback to DVDs. For some people, Macrovision copy protection (refer to "What Are the Copy Protection

Issues?" in Chapter 1) is an annoyance. Laserdiscs have no copy protection and do not have regional differences other than PAL versus
NTSC.

• **Recordable** DVD recorders are increasingly affordable. Laserdisc
recording, at a low of $250 per disc, was never available to general
consumers.

For more laserdisc info, see Leopold's FAQ at www.cs.tut.fi/,leopold/
Ld/FAQ/index.html and Bob Niland's FAQs and overview at www.access-
one.com/rjn/laser/laserdisc.html (overview reprinted from *Widescreen
Review Magazine*).

Can I Modify or Upgrade My Laserdisc Player to Play DVDs?

No, DVD circuitry is completely different, the pickup laser is a different
wavelength, and the tracking control is more precise. No hardware
upgrades have been announced, and in any case they would be more
expensive than buying a DVD player to put next to the laserdisc player.

Do DVDs Support HDTV (DTV)? Will HDTV Make DVDs Obsolete?

The short answers are partially and no.

First, some quick definitions: HDTV encompasses both analog and digital TVs that have a 16:9 aspect ratio and approximately five times the resolution of standard TV (double vertical, double horizontal, and a wider
aspect). *Digital TV* (DTV) applies to digital broadcasts in general and to the
U.S. *Advanced Television Systems Committee* (ATSC) standard in specific.
The ATSC standard includes both *standard-definition* (SD) and *high-
definition* (HD) digital formats. The notation H/DTV is often used to specifically refer to high-definition DTV.

In December of 1996, the *Federal Communications Commission* (FCC)
approved the U.S. DTV standard. HDTVs became available in late 1998, but
they are expensive and won't become widespread for many years. DVDs
are not HD, but they look great on HDTVs. Over 80 percent of the 2 million
DTV sets sold in the United States in 2002 did not have tuners, indicating
that their owners got them for watching DVDs.

DVD-Video does not directly support HDTV. No digital HDTV standards
were finalized when DVD was developed. In order to be compatible with
existing televisions, DVD's MPEG-2 video resolutions and frame rates are
closely tied to NTSC and PAL, SECAM video formats (see Chapter 1's "Is

DVD-Video a Worldwide Standard? Does It Work with NTSC, PAL, and SECAM?"). DVD does use the same 16:9 aspect ratio of HDTV and the Dolby Digital audio format of U.S. DTV.

HDTV in the United States is part of the ATSC DTV format. The resolution and frame rates of DTV generally correspond to the ATSC recommendations for SD (640×480 and 704×480 at 24p, 30p, 60p, and 60i) and HD (1280×720 at 24p, 30p, and 60p; 1920×1080 at 24p, 30p, and 60i). (24p means 24 progressive *frames per second* [fps], 60i means 60 interlaced fields per second [30 fps].) The current DVD-Video spec covers all of SD except 60p. It's expected that future DVD players will output digital video signals from existing discs in SDTV formats. The HD formats are 2.7 and 6 times the resolution of DVD, and the 60p version is twice the frame rate. The *International Telecommunication Union – Radio Communication Sector* (ITU-R) is working on BT.709 HDTV standards of 1125/60 ($1920 \times 1035/30$) (the same as SMPTE 240M, similar to Japan's analog MUSE HDTV) and 1250/50 ($1920 \times 1152/25$), which may be used in Europe. The latter is 5.3 times the resolution of DVD's $720 \times 576/25$ format. The HD maximum data rate is usually 19.4 Mbps, almost twice the maximum DVD-Video data rate. In other words, DVD-Video does not currently support HDTV video content.

HDTV will not make DVDs obsolete. Those who postpone purchasing a DVD player because of HDTV are in for a long wait. HDTVs became available in late 1998 at very high prices (about $5000 and up). It will take many years before even a small percentage of homes have HDTV sets. The CEA expects 10 percent of U.S. households to have HDTV in 2003, 20 percent by 2005, and 30 percent by 2006.

HDTV sets include analog video connectors (composite, s-video, and component) that work with all DVD players and other existing video equipment such as VCRs. Existing DVD players and discs will work perfectly with HDTV sets and provide a much better picture than any other prerecorded consumer video format, especially when using a progressive-scan player. Because the cheapest route to HDTV reception will be HDTV converters for existing TV sets, broadcast HDTV for many viewers will look no better than DVDs.

HDTV displays support digital connections such as HDMI (DVI) and IEEE 1394/FireWire, although standardization is not quite finished. Digital connections for audio and video provide the best possible reproduction of DVDs, especially in widescreen mode. The DVD Forum finalized specifications for supporting 1394 and HDMI in 2002 and players ith DVI/HDMI digital outputs appared in 2003. When the DVD *stream recording* (SR) format is finalized, DVD-SR players may be usable as "transports" that output any kind of audio/video data (even formats developed after the player was built) to different sorts of external displays or converters.

The interesting thing many people don't realize is that DTV happened sooner, faster, and cheaper on PCs. A year before any consumer DTV sets came out, you could buy a DVD PC with a 34-inch VGA monitor and get gorgeous progressive-scan movies for under $3000. The quality of a good DVD PC connected to a data-grade video projector can beat a $30,000 line-doubler system. (See BroadbandMagic, Digital Connection, and Sleak-line for product examples. Video projectors are available from Barco, Dwin, Electrohome, Faroudja, InFocus, Projectavision, Runco, Sharp, Sony, Vidikron, and others.)

Eventually, the DVD-Video format will be upgraded to an HD-DVD format. See "Will High-Definition DVDs or 720p DVDs Make Current Players and Discs Obsolete?" later in this chapter. Also see "What's New with DVD Technology?" in Chapter 6, "Miscellaneous."

What Is Divx?

Two forms of Divx exist. The original was a pay-per-view version of DVD. The later version is a video encoding format.

The Original Divx

Depending on whom you ask, Divx (Digital Video Express first known as ZoomTV) was either an insidious evil scheme for greedy studios to control what you see in your own living room or an innovative approach to video rental that would offer cheap discs you could get almost anywhere and keep for later viewings.

Developed by Circuit City and a Hollywood law firm, Divx was supported by Disney (Buena Vista), Twentieth Century Fox, Paramount, Universal, MGM, and DreamWorks SKG, all of which also released discs in the "open DVD" format, because the Divx agreement was nonexclusive. Harman/Kardon, JVC, Kenwood, Matsushita (Panasonic), Pioneer, Thomson (RCA/Proscan/GE), and Zenith announced Divx players, although some never came to market (The ones that did were Panasonic X410, Proscan PS8680Z, RCA RC5230Z and RC5231Z, and Zenith DVX2100). The studios and hardware makers supporting Divx were given incentives in the form of guaranteed licensing payments totaling over $110 million. Divx discs were manufactured by Nimbus, Panasonic, and Pioneer. Circuit City lost over $114 million (after tax write-offs) on Divx.

As stated, Divx was a pay-per-viewing-period variation of DVD. Divx discs sold for $4.50. Once inserted into a Divx player, the disc would play normally (allowing the viewer to pause, rewind, or even put in another disc before finishing the first one) for the next 48 hours, after which the owner had to pay $3.25 to unlock it for another 48 hours. A Divx DVD player, which cost about

$100 more than a regular player, had to be hooked up to a phone line so it could call an 800 number for about 20 seconds during the night once each month (or after playing 10 or so discs) to upload billing information.

Most Divx discs could be converted to DivxSilver status by paying an additional fee (usually $20) to allow unlimited plays on a single account (as of December 1998, 85 percent of Divx discs were convertible). Unlimited-playback DivxGold discs were announced but never produced.

Divx players can also play regular DVD discs, but Divx discs do not play in standard DVD players. Divx discs are serialized with a barcode in the standard burst cutting area. In addition to normal DVD copy protection (refer to "What Are the Copy Protection Issues?" in Chapter 1), they employ watermarking the video, modified channel modulation, and triple *Data Encryption Standard* (DES) encryption (two 56-bit keys) of serial communications. Divx technology never worked on PCs, which undoubtedly contributed to its demise. Because of the DES encryption, Divx technology may not have been allowed outside the United States.

Divx was originally announced for summer 1998 release. Limited trials began on June 8, 1998, in San Francisco and in Richmond, Virginia. The only available player was from Zenith (which at the time was in Chapter 11 bankruptcy), and the promised 150 movies had dwindled to 14. The limited nationwide rollout (with one Zenith player model and 150 movies in 190 stores) began on September 25, 1998. By the end of 1998, about 87,000 Divx players (from four models available) and 535,000 Divx discs were sold (from about 300 titles available). The company apparently counted the 5 discs bundled with each player, which means 100,000 additional discs were sold. By March 1999, 420 Divx titles were available (compared to over 3,500 open DVD titles). All things considered, Divx players were selling well and titles were being produced with impressive speed.

On June 16, 1999, less than a year after initial product trials, Circuit City withdrew its support and Divx announced it was closing down. Divx did not confuse or delay development of the DVD market nearly as much as many people predicted (including yours truly). In fact, it probably helped by stimulating Internet rental companies to provide better services and prices, by encouraging manufacturers to offer more free discs with player purchases, and by motivating studios to develop rental programs.

When it closed down, the company offered $100 rebate coupons to all owners of Divx players. This made the players a good deal, because they could play open DVDs just as well as other low-end players that cost more. On July 7, 2001, Divx players dialed into the central billing computer, which decommissioned them. (Divx players not connected to phone lines have expired their playback allowance.) Consequently, Divx discs are no longer playable in any players. For more information see the Divx Owner's Association (www.the-doa.com).

The advantages of Divx were as follows:

- Viewing could be delayed, unlike rentals.

- Discs need not be returned. No late fees.

- You could watch the movie again for a small fee. Initial cost of "owning" a disc was reduced.

- Discs could be unlocked for unlimited viewing (DivxSilver), an inexpensive way to preview before deciding to purchase.

- The disc was new, with no damage from previous renters.

- The rental market was opened up to other retailers, including mail order.

- Studios got more control over the use of their content.

- You received special offers from studios in your Divx mailbox.

- Divx players (with better quality and features than comparable players) were a steal after Divx went out of business.

The disadvantages of Divx were as follows:

- The player cost was higher than a DVD player (about $100 more at first, about $50 later).

- Although discs did not have to be returned, the viewer still had to go through the effort of purchasing the disc. Cable/satellite pay per view is more convenient.

- Divx cost more than a regular DVD rental ($3 to $7 versus $2 to $4). The company raising prices later had few obstacles, because it had a monopoly.

- Casual quick viewing (looking for a name in the credits, playing a favorite scene, or watching supplements) required paying a fee.

- Most Divx titles were pan and scan (see "What's Widescreen? How Do the Aspect Ratios Work?" in Chapter 3) without extras such as foreign language tracks, subtitles, biographies, trailers, and commentaries.

- The player had to be hooked to your phone line, possibly requiring a new jack in your living room or a phone extension cable strung across it. Players required a connection once a month or so, so you could periodically connect it to a phone line.

- Divx couldn't be used in mobile environments, such as a van or RV, unless you took it out and connected it to a phone line about once a month.

- The Divx central computer collected information about your viewing habits, as do cable/satellite pay-per-view services and large rental chains. According to Divx, the law did not allow them to use the information for resale and marketing.
- Divx players included a mailbox for companies to send you unsolicited offers (spam).
- Those who didn't lock out their Divx player could receive unexpected bills when their kids or visitors played Divx discs.
- Divx discs wouldn't play in regular DVD players or on PCs with DVD-ROM drives. Some uninformed consumers bought Divx discs only to find they wouldn't play in their non-Divx player.
- Unlocked Silver discs would only work in players on the same account. Playback in a friend's Divx player would incur a charge. (Gold discs, which were never released, would have played without charge in all Divx players.)
- No market existed for used Divx discs.
- Divx discs became unplayable after June 2001.
- Divx players were never available outside the United States and Canada.

The New DivX

In March of 2000, a DVD redistribution technology called DivX;-) appeared. Yes, the smiley face was originally part of the name, which was a take-off on the original Divx format. The perpetrators should be drawn and quartered for the stupid joke, which caused untold confusion.

DivX was originally a simple hack of Microsoft's MPEG-4 video codec, combined with MP3 audio, allowing decrypted video from a DVD to be reencoded for downloading and playing in Windows Media Player. Work on DivX evolved through Project Mayo and a version originally called DivX Deux into an open-source initiative known as OpenDivX, based on the MPEG-4 standard. Out of all this came DivXNetworks, a company that turned DivX into an extensive video encoding and delivery system. An open-source variation is called 3ivx.

How Can I Record from DVD to Videotape?

Why in the world would you want to degrade a DVD's beautiful digital picture by copying it to analog tape, especially because you lose the interactive menus and other nice features?

If you really want to copy to VHS, hook the audio/video outputs of the DVD player to the audio/video inputs of your VCR, and then record the disc to tape. You'll discover that most of the time the resulting tape is garbled and unwatchable. This is because of the Macrovision feature designed to prevent you from doing this. Refer to "What Are the Copy Protection Issues?" in Chapter 1.

Will High-Definition DVD or 720p DVD Make Current Players and Discs Obsolete?

Not for a long time. HD-DVD is just becoming available. HD stands for both *high density* (more data on a disc) and *high definition* (better quality picture). The first commercial Blu-ray HD-DVD recorders appeared in Japan in April of 2003, over seven years after DVD was introduced there. The recorders are designed for home recording only (not for playing prerecorded HD movies), and only work with Japan's digital HD broadcast system.

New DVD formats will slowly supercede the original DVD format, but new players will play old DVD discs and often make them look even better (with progressive-scan video and picture processing). However, new HD-DVD discs won't be playable in old DVD players (unless they are special hybrid discs in both HD and SD format). Your collection of standard DVDs will be playable for many years to come, and titles will only become obsolete in the sense that you might want to replace them with new high-definition versions. Consider that U.S. HDTV was anticipated to be available in 1989, yet it was not finalized until 1996 and did not appear until 1998. Has it made your current TV obsolete yet? See "What About the HD-DVD and Blue Laser Formats?" for details on HD-DVD, and *see* Chapter 6 for more information on the future of DVD.

Ironically, computers supported HDTV before set-top players, because double-headed DVD-ROM drives with appropriate playback and display hardware meet the 19 Mbps data rate needed for HDTV. This led to various 720p DVD projects that use the existing DVD format to store video in 1280×720 or 1920×1080 resolution at 24 progressive fps. It's possible that 720p DVDs can be made compatible with existing players (which would only recognize and play the 480-line line data).

NOTE: The term *HDVD* has already been taken to mean high-density volumetric display (see www.3dmedia.com).

Some have speculated that a double-headed player reading both sides of the disc at the same time could double the data rate or provide an enhancement for applications such as HDTV. This is currently impossible because the track spirals go in opposite directions (unless all four layers are used). The DVD spec would have to be changed to allow reverse spirals on layer 0. Even then, keeping both sides in sync, especially with MPEG-2's variable bit rate, would require independently tracking heads, precise track and pit spacing, and a larger, more sophisticated track buffer. Another option would be to use two heads to read both layers of one side simultaneously. This is technically feasible but has no advantage over reading one layer twice as fast, which is simpler and cheaper. (Refer to "Do DVDs Support HDTV (DTV)? Will HDTV Make DVDs Obsolete?" for more information about HDTV and DVDs.)

What Effect Will Fluorescent Multilayer Disc (FMD) Have on DVD?

Very little, as Constellation 3D ran out of money in mid-2002. The various reports of *fluorescent multilayer disc* (FMD) causing the early death of DVD were wildly exaggerated and not founded in reality.

Fluorescent multilayer technology, which can be used in cards or discs, aims a laser at fluorescent dye, causing it to emit light. Because it doesn't depend on reflected laser light, it's possible to create many data layers (C3D prototyped 50 layers in its lab). It can use the same 650-nanometer laser as DVDs, so FMD drives could be made to read DVDs. In June of 2000, C3D announced a program to make FMDs with 25 GB per side that would be readable by DVD drives with a "minor and inexpensive modification." C3D later said players would be available by mid-2001.

FMD was very cool technology, but it was new, with no track record, developed by one small company. DVD is based on decades of optical storage technology development by dozens of companies. The monumental task of changing entire production infrastructures over to a new format was too much for C3D, even with tens of millions of dollars and some large partners.

How Does MPEG-4 Affect DVD?

MPEG-4 is a video encoding standard designed primarily for low-data rate streaming video, although it's actually more efficient than MPEG-2 at DVD and HDTV data rates. MPEG-4 provides for advanced multimedia with *media objects*, but most implementations only support simple video (*Simple Visual Profile*). MPEG-4 part 10, also known as H.264 (and also known as JVT or AVC) is an even better encoding standard.

DVDs use MPEG-2 video encoding (see "What Are the Video Details?" in Chapter 3), and standard DVD players don't recognize the MPEG-4 video format. MPEG-4 files, however, can be stored on DVD-ROMs for use on computers. For example, DivX uses MPEG-4 (see "The NewDivX" earlier in this chapter).

It's possible that MPEG-4 or H.264 will be used in a future, high-definition version of DVD. In any case, it will probably not appear before 2005 at the earliest. For more about MPEG, see Tristan's MPEG.org site and the MPEG home page (www.cselt.it/mpeg).

What's WebDVD or Enhanced DVD?

WebDVD is the simple but powerful concept of combining DVD content with Internet technology. It combines the best of DVD (fast access to high-quality video, audio, and data) with the best of the Internet (interactivity, dynamic updates, and communication). In general, WebDVD refers to enhancing a DVD with HTML pages and links, or enhancing a web site with content from a local DVD drive.

WebDVD is not a trademarked term of AOL/Time Warner, Microsoft, or any other company. Variations on the WebDVD concept are known as iDVD, eDVD, Connected DVD, and so on. It's not a new idea; it's been done on CD-ROMs for years, but the differences with DVD are that the quality of the audio and video are finally better than TV, and the discs can be played in low-cost set-top players.

Almost all WebDVD implementations are currently available for PCs, but new players are adding WebDVD features. A working group of the DVD Forum is creating a standardized Enhanced DVD format for set-top DVD players.

Most major authoring systems (see "Which DVD Authoring Systems Are Available?" in Chapter 5) include rudimentary tools for adding HTML enhancements to DVD. For fancier WebDVD development, a variety of tools are available (see "How Do I Play DVD-Video in HTML, PowerPoint, Director, VB, and So On?" in Chapter 4).

For more on WebDVD, see Phil DeLancie's EMedia article, "Untangling Web-DVD Playback" (www.emedialive.com/r8/2001/delancie2_01.html). Good examples of WebDVD sites are *Mars: The Red Planet, Stargaze*, and *DVD Demystified*. The authors of these sites (Ralph LaBarge and Jim Taylor) encourage you to copy their code as a starting place for your own Web-DVD creations.

What's a Nuon Player?

Nuon was a specialized "media processor" chip designed by VM Labs that was powerful enough to play DVDs and video games. The chip was originally intended for video game consoles, but it was hitched to the DVD wagon when the game market dried up and the DVD market exploded. Some DVD players from Samsung, Thomson (RCA), and Toshiba were built on Nuon technology. The extra processing power in a Nuon player enabled special features such as graphical overlays, digital zoom, and live thumbnails. Some DVD movies were produced with added content designed specifically for the Nuon platform. As of the beginning of 2002, four Nuon-enhanced DVD movies were on the market: *The Adventures of Buckaroo Banzai Special Edition, Bedazzled, Dr. Doolittle 2*, and *Planet of the Apes*.

In December 2001, VM Labs filed for Chapter 11 bankruptcy, and in March 2002 the company's assets were purchased by Genesis Microchip. A new division, Nuon Semiconductor, was formed to market Nuon chips under the Aries name. On July 24, 2002, Genesis laid off the entire Nuon division.

Chapter 3

DVD Technical Details

What Are the Outputs of a DVD Player?

DVD players usually have two or three kinds of video output (composite, s-video, and component) and three or four kinds of audio output (analog stereo, digital *Pulse Code Modulation* [PCM] stereo, Dolby Digital, and *Digital Theater Systems* [DTS]). More details will be covered in this chapter, particularly in the "How Do I Hookup a DVD Player?" section.

Video Outputs

Most DVD players have the following video output connections, which can carry an NTSC, PAL, or SECAM signal:

- **S-video (Y/C)**   A four-pin round plug that carries a brightness signal (Y) and two combined color signals (C).

- **Composite video (CVBS)**   A standard yellow RCA video plug that combines all three video signals into one.

European players combine both of the signals above, along with others, into a 21-pin rectangular SCART connector (also known as a Peritel or *Euro Connector* [EC]).

Some players may have additional video connections:

- Component-interlaced analog video (EIA 770.1) that keeps all three video signals separate. This comes in two different formats:

  - **Y'PbPr format**   Three RCA connectors or BNCs connectors.

  - **Red Green Blue (RGB), RGB Composite Sync (RGBS), or RGB Horizontal Sync Vertical Sync (RGBHV) format**   SCART connector or 3, 4, or 5 RCA or BNC connectors.

- Component progressive analog video that keeps all three video signals separate. It comes in Y'PbPr format with 3 RCA connectors, or RGB (or RGBS or RGBHV) formats with 3, 4, or 5 RCA or BNC connectors.

- *Radio-frequency* (RF) video for connecting to the TV antenna input, usually on channel 3 or 4 using a screw-on, 75-ohm, F-type connector. It may require an adapter for TVs that have 300-ohm, two-screw, antenna wire connectors.

- *High-definition multimedia interface* (HDMI), which is digital video in the *Digital Visual Interface* (DVI) format, plus digital audio.

Most DVD players with component video outputs use YUV (Y'PbPr), which is incompatible with RGB equipment. European players with component video outputs usually provide RGBS signals on the SCART connector. YUV-to-RGB transcoders are rumored to be available for $200 to $300, but seem hard to track down. A $700 converter is available from avscience, and a $900 converter, the CVC 100, is available from Extron. Converters are also available from Altinex, Kramer, Monster Cable, and others. For progressive scan, you need a converter that can handle 31.5 kHz signals. Converters from s-video are also an option (available from Markertek and others).

NOTE: The correct term for analog color-difference output is Y'Pb'Pr', not Y'Cb'Cr' (which is digital, not analog). To simplify things, this book sometimes uses the term YUV in the generic sense to refer to analog color difference signals.

Specialty players from companies such as Function Communications, Theta Digital, and Vigatec are available with *serial digital interface* (SDI) output, but they connect only to high-end or production equipment.

Audio Outputs

Most DVD players have the following audio output connections:

- Analog stereo audio, which may be in Dolby Surround, depending on the disc.
  - Two RCA connectors, red and white.
  - European players transmit analog stereo audio on the SCART connector.

- Digital audio with 1 to 5.1 channels: raw digital audio in PCM, MLP, Dolby Digital (AC-3), DTS, or MPEG-2. It requires an amplifier/receiver with a built-in decoder (or a separate external decoder). These audio connectors come in two different formats:

  - S/PDIF coax format, as an RCA connector (IEC-958 Type II).

  - Toslink format, as a square optical connector (EIAJ CP-340 and EIAJ CP-1201).

Some players may have additional audio connections:

- Multichannel analog audio that requires a multichannel-ready or Dolby Digital-ready amplifier/receiver with six inputs. Provided on six RCA connectors or one DB-25 connector.

- AC-3 RF audio, but only on combination laserdisc/DVD players. This carries audio from AC-3 laserdiscs and uses one RCA connector.

- High-resolution digital audio in one of two formats:

  - 1394 (FireWire), which is a rectangular connector that requires a receiver with a 1394 audio input.

  - HDMI that requires a receiver/TV with HDMI input.

Some players and receivers support only S/PDIF or only Toslink. If your player and receiver don't match, you'll need a converter, such as the Audio Authority 977Midiman C02, COP 1, or POF.

Some players can output 96/24 PCM audio using a nonstandard variation of IEC-958 running at 6.144 Mbps instead of the normal limit of 3.1 MHz. Note that the *Content Scrambling System* (CSS) license does not allow digital PCM output of CSS-protected material at 96 kHz. The player must downsample to 48 kHz. The *Pioneer Elite DV-47Ai* is the only DVD player (as of September 2002) with DTCP-protected 1394 output for full, multichannel 96/24 and 192/24 PCM.

How Do I Hook up a DVD Player?

It depends on your audio/video system and your DVD player. Most DVD players have two or three video hookup options and three audio hookup options. Choose the output format with the best quality (as indicated) that is supported by your video and audio systems. See "What Are the Outputs of a DVD Player?" earlier in this chapter, for output connector details.

On many TVs, you will need to switch the TV to auxiliary input (line input). You might need to tune it to channel 0 to make this work.

If you want to hook multiple devices (DVD player, VCR, cable/satellite box, and WebTV) to a single TV, you will need one of the following:

- A TV with multiple inputs.

- A manual audio/video switchbox ($30 at electronics suppliers such as Comtrad).

- An audio/video receiver (to switch between video sources via remote control). If you plan on getting an audio/video receiver, make sure it can handle the video format you want to use (component or s-video).

Video Hookup

The following video hookup options are available:

- **S-video (good quality)**   Almost all DVD players have s-video output, which looks much better than composite video and is only slightly inferior to component video. Hook an s-video cable from the player to the display (or to an audio/video receiver that can switch s-video). The round, four-pin connector may be labeled Y/C, s-video, or S-VHS. If you're in Europe, you can use a SCART cable.

- **Composite video (okay quality)**   All DVD players have standard RCA (Cinch) baseband video connectors or a SCART connector (in Europe) to carry composite video. Hook a standard video cable or a SCART cable from the player to the display (or to an audio/video receiver to switch the video). The RCA video connectors are usually yellow and may be labeled video, CVBS, composite, or baseband.

- **Component video (better quality)**   Some U.S. and Japanese players have interlaced component YUV (Y'Pb'Pr') video output. Connectors may be labeled YUV, color difference, YPbPr, or Y/B-Y/R-Y, and they may be colored green/blue/red. (Some players incorrectly label the output as YCbCr.) Some players have RGB component video output via a SCART connector or three RCA or BNC connectors labeled R/G/B. Hook cables from the three video outputs of the player to the three video inputs of the display, or hook a SCART cable from the player to the display.

NOTE: No standardization on the output interface format exists (voltage and setup). Players apparently use SMPTE 253M (286 mV sync, 0 percent luma setup with 700 mV peak, and ±300 mV color excursion), Betacam (286 mV sync, 7.5 percent luma setup with 714

mV peak, and ±350 mV color excursion), M-II (300 mV sync, 7.5 percent luma setup with 700 mV peak, and ±324.5 mV color excursion),
or nonstandard variations. Note that outputs with no IRE setup can
provide a wider range of luma values for a slightly better picture. For
equipment with an RGB input, a YUV converter is usually needed.
(Refer to "What Are the Outputs of a DVD Player?")

- **Progressive video (even better quality)**   A few players have progressive-scan YUV (Y'Pb'Pr') or RGB (European players only) component video output. Hook decent-quality cables from the three video
  outputs of the player to the three video inputs of a progressive-scan
  line multiplier or a progressive-scan TV. You can also use a SCART
  cable if you have a European player and progressive-scan TV with the
  right connectors. Toshiba calls progressive scan ColorStream PRO.
  Progressive video preserves the progressive nature of most movies,
  providing a film-like, flicker-free image with improved vertical resolution and smoother motion. DVD computers can also produce progressive video from DVD. In this case, use a 15-pin computer video
  cable to connect the VGA output of the PC to the VGA input of a monitor or projector. If the projector only has RGB or YPbPr inputs, you'll
  need a converter such as the Audio Authority 9A60. (Refer to "What's
  a Progressive DVD Player?" in Chapter 1, "General DVD," and
  "Will High-Definition DVDs or 720p DVDs Make Current Players and
  Discs Obsolete?" in Chapter 2, "DVD's Relationship to Other Products
  and Technologies" for more information on progressive video. Also
  see "Can I Play Movies on My Computer?" in Chapter 4, "DVDs and
  Computers.")

- **Digital video (best quality)**   A few players have HDMI (DVI) or 1394
  digital outputs. This preserves the true digital signal from the DVD.
  Hook an HDMI or 1394 cable from the output of the player to the HDMI
  or 1394 input of a digital television or other digital audio/video system.
  The same cable carries the digital audio signal.

- **RF video (worst quality)**   You should use this connection only if you
  have an old TV that has only a screw-on antenna input. Most DVD
  players don't have RF output, so you will probably need to buy an RF
  modulator ($30 at Radio Shack, Comtrad, or Markertek). But first see
  the following warning about using a VCR as an RF modulator. If the
  player has built-in RF output, it will include audio, although it may only
  be mono.

  To set everything up, connect a coax cable from the yellow video output of the player to the input of the modulator. If you are not hooking the

player up to a separate stereo system, connect a coax cable from the left audio output of the player to the audio input of the modulator. (If you have a stereo modulator, connect another cable for the right audio channel.)

Then connect a coax antenna cable from the modulator to the TV. You may need a 300-ohm to 75-ohm adapter (to switch between a two-wire antenna connection and a threaded coax connection). Tune the TV to channel 3 or 4 (or channel 36 in Germany and some other European countries) and set the switch on the modulator or the back of the player to match. If you also want to hook up a VCR, connect an antenna cable from the output of the VCR to the antenna input of the modulator.

WARNING: If you connect your DVD player to a VCR and then to your TV (or to a combination TV/VCR), you will probably have problems with discs that enable the player's Macrovision circuit. (See "Will I Have Problems Connecting My VCR Between My TV and My DVD Player?")

Also, some video projectors don't recognize the 4.43 NTSC signal from NTSC discs in PAL players (see "Is DVD-Video a Worldwide Standard? Does It Work with NTSC, PAL, and SECAM?" in Chapter 1). They see the 60 Hz scanning frequency and switch to NSTC, even though the color subcarrier is in PAL format.

NOTE: Most DVD players support widescreen signaling, which tells a widescreen display what the aspect ratio is so that it can automatically adjust. One standard (*International Telecommunication Union-Radio Communication Sector* [ITU-R] BT.1119, used mostly in Europe) includes information in a video scanline. Another standard, for Y/C connectors, adds a 5V DC signal to the chroma line to designate a widescreen signal. Unfortunately, some switchers and amps throw away the DC component instead of passing it on to the TV.

For more information on conversions between formats, see the amazing Notes on Video Conversion from the Sci.Electronics.Repair FAQ at www.repairfaq.org/sam/vodconv.htm.

Audio Hookup

The following audio hookup options are available:

NOTE: All DVD players have a built-in two-channel Dolby Digital (AC-3) decoder. Some can also decode MPEG or DTS audio. The decoder translates multichannel audio into two-channel PCM audio. This goes to the digital output and is also converted to analog for standard audio output. Some players have a built-in multichannel Dolby Digital decoder, but it's only useful if you have an audio system with multichannel analog inputs. (See "Can You Explain This Dolby Digital, Dolby Surround, Dolby Pro Logic, DTS Stuff in Plain English?" for more explanation.)

- **Analog audio (two-channel stereo/Surround) (ok quality)**   All DVD players include two RCA connectors for stereo output. Any disc with multichannel audio is automatically decoded and downmixed to Dolby Surround output for connection to a regular stereo system or a Dolby Surround/Pro Logic system. Connect two audio cables between the player and receiver, amplifier, or TV. Connectors may be labeled audio or left/right; left is usually white, and right is usually red. If your TV has only one audio input, use the left channel from the DVD player.

- **Digital audio (best quality)**   Almost all DVD players have digital audio outputs. The same output can carry Dolby Digital (AC-3), PCM audio (including PCM from CDs), DTS, MPEG-2 audio (PAL/SECAM players only), and MLP audio (from DVD-Audio discs). For PCM, a digital receiver or an outboard DAC is required. For all other formats, the appropriate decoder is required in the receiver/amplifier or as a separate audio processor.

For example, to play a disc with a Dolby Digital soundtrack using a digital audio connection, the receiver has to have the Dolby Digital feature. DTS discs require a player with the "DTS Digital Out" mark (older players don't recognize DTS tracks) and the DTS decoding feature in the receiver. (All DVD players can play DTS CDs if a DTS decoder is connected to the digital PCM output signal.) Some DVD players have coax connectors (SP/DIF), some have fiber-optic connectors (Toslink), and many have both. Endless arguments take place over which of these is better. Coax seems to have more advocates, because it's inherently simpler. Optical cable is not affected by electromagnetic interference, but it's more fragile and can't curve tightly. Suffice it to say that because the signal is digital, a quality cable of either type will provide similar results. Hook a 75-ohm coax cable or a fiber-optic cable between the player and the receiver. You might need a converter; refer to "What Are the Outputs of a DVD Player?"

Some players provide separate connectors for Dolby Digital/DTS/MPEG and for PCM. On others, you may need to select the desired output format using the player setup menu or a switch on the back of the player. If you try to feed Dolby Digital or DTS to a digital receiver that doesn't recognize it, you'll get no audio.

NOTE: Make sure you use a quality cable; a cheap RCA patch cable may cause the audio to sound poor or not work at all. Also, connecting to the AC-3/RF (laserdisc) input of a receiver will not work unless your receiver can autoswitch, because DVD digital audio is not in RF format (see the end of this list for info).

- **Component analog audio (excellent quality)** Some players provide six-channel analog output from the internal Dolby Digital or DTS decoder. A few provide seven-channel output from 6.1 tracks. The digital-to-analog conversion quality in the player may be better or worse than in an external decoder. A receiver/amplifier with six or seven inputs (or more than one amplifier) is required; this type of unit is often called Dolby Digital-ready or AC-3-ready. Unfortunately, in many cases you won't be able to adjust the volume of individual channels or perform bass management. Hook six (or seven) audio cables to the RCA connectors on the player and to the matching connectors on the receiver/amplifier. Some receivers require an adapter cable with a DB-25 connector on one end and RCA connectors on the other.

NOTE: Until a digital connection standard is created, the only way to get multichannel PCM output from DVD-Audio players will be with analog connections or proprietary connections. If you plan to get a DVD-Audio player, you'll need a receiver with analog multichannel inputs.

- **RF digital audio (laserdisc only)** Combination laserdisc/DVD players include AC-3 RF output for digital audio from laserdiscs. Hook a coax cable to the AC-3 RF input of the receiver/processor. Note that digital audio from DVDs does not come out of the RF output; it comes out of the optical/coax outputs. Analog audio from LDs will come out the stereo connectors, so three separate audio hookups are required to cover all variations.

Will I Have Problems Connecting My VCR Between My TV and My DVD Player?

It's not a good idea to route the video from your DVD player through your VCR. Most movies use Macrovision protection (see "What Are the Copy Protection Issues" in Chapter 1), which affects VCRs and causes problems such as a repeated darkening and lightening of the picture. If your TV doesn't have a direct video input, you may need a separate RF converter (refer to "How Do I Hook up a DVD Player?" in this chapter) or, better yet, get a new TV with direct video inputs.

You may also have problems with a TV/VCR combo, because many of them route the video input through the VCR circuitry. The best solution is to get a box to strip Macrovision.

Why Is the Audio or Video Bad?

The number one cause of bad video is a poorly adjusted TV. The high fidelity of DVD-Video demands much more from the display. Turn the sharpness and brightness down. Refer to Chapter 1's "What's the Quality of DVD-Video?" for more information. For technical details of TV calibration, see Anthony Haukap's FAQ: How to Adjust a TV.

If you get audio hum or noisy video, it's probably caused by interference or a ground loop. Try a shorter cable. Make sure the cables are adequately shielded. Try turning off all equipment except the pieces you are testing. Move things farther apart. Try plugging into a different circuit. Make sure all equipment is plugged into the same outlet. Ground your braces. Wrap your entire house in tinfoil. For more on ground loops, see www.hut.fi/Misc/Electronics/docs/groundloop/. More information for repair technicians is available at shophelper.net.

If the video freezes or breaks up, it may be caused by scratches on the disc (refer to "How Should I Clean and Care for DVDs?" in Chapter 1). It's normal for DVDs to freeze for a fraction of a second in the middle of a movie; this is a layer break (refer to "What Is a Layer Change? Where Is It on Specific Discs?" in Chapter 1).

What Are the Sizes and Capacities of DVD?

Many variations on the DVD theme have been created. Two physical sizes are available: 12 centimeter (4.7 inches) and 8 centimeter (3.1 inches), both

1.2 millimeters thick, made of two 0.6-millimeter substrates glued together. These are the same form factors as a CD. A DVD can be single-sided or double-sided. Each side can have one or two layers of data. The amount of video a disc can hold depends on how much audio accompanies it and how heavily the video and audio are compressed. The oft-quoted figure of 133 minutes is apocryphal; a DVD with only one audio track easily holds over 160 minutes, and a single layer can actually hold up to 9 hours of video and audio if it's compressed to VHS quality.

At a rough average rate of 5 Mbps (4 Mbps for video and 1 Mbps for 2 or 3 tracks or audio), a single-layer DVD can hold a little over 2 hours. A dual-layer disc can hold a 2-hour movie at an average of 9.5 Mbps (close to the 10.08 Mbps limit).

A DVD-Video disc containing mostly audio can play for 13 hours (24 hours with dual layers) using 48/16 PCM (slightly better than CD quality). It can play 160 hours of audio (or a whopping 295 hours with dual layers) using Dolby Digital 64 kbps compression of monophonic audio, which is perfect for audio books.

DVD Capacities

For reference, a CD-ROM holds about 650 MB, which is 0.64 gigabytes or 0.68 billion bytes. Table 3-1 outlines the capacities of all the different versions of DVDs. SS/DS means *single-sided/double-sided*, SL/DL/ML means *single-layer/dual-layer/mixed-layer* (mixed means single layer on one side and dual layer on the other side), gig means gigabytes (2^{30}) and BB means billions of bytes (10^9). See note about giga versus billion in "Notation and Units" in Chapter 7, "Leftovers."

TABLE 3-1   Various DVD Capacities

DVD-5 (12 cm, SS/SL)	4.37 gig (4.70 BB) of data, over 2 hours of video
DVD-9 (12 cm, SS/DL)	7.95 gig (8.54 BB), about 4 hours
DVD-10 (12 cm, DS/SL)	8.74 gig (9.40 BB), about 4.5 hours
DVD-14 (12 cm, DS/ML)	12.32 gig (13.24 BB), about 6.5 hours
DVD-18 (12 cm, DS/DL)	15.90 gig (17.08 BB), over 8 hours
DVD-1 (8 cm, SS/SL)	1.36 gig (1.46 BB), about half an hour
DVD-2 (8 cm, SS/DL)	2.47 gig (2.66 BB), about 1.3 hours
DVD-3 (8 cm, DS/SL)	2.72 gig (2.92 BB), about 1.4 hours
DVD-4 (8 cm, DS/DL)	4.95 gig (5.32 BB), about 2.5 hours

DVD-R 1.0 (12 cm, SS/SL)	3.68 gig (3.95 BB)
DVD-R 2.0 (12 cm, SS/SL)	4.37 gig (4.70 BB)
DVD-R 2.0 (12 cm, DS/SL)	8.75 gig (9.40 BB)
DVD-RW 2.0 (12 cm, SS/SL)	4.37 gig (4.70 BB)
DVD-RW 2.0 (12 cm, DS/SL)	8.75 gig (9.40 BB)
DVD+R 2.0 (12 cm, SS/SL)	4.37 gig (4.70 BB)
DVD+R 2.0 (12 cm, DS/SL)	8.75 gig (9.40 BB)
DVD+RW 2.0 (12 cm, SS/SL)	4.37 gig (4.70 BB)
DVD+RW 2.0 (12 cm, DS/SL)	8.75 gig (9.40 BB)
DVD-RAM 1.0 (12 cm, SS/SL)	2.40 gig (2.58 BB)
DVD-RAM 1.0 (12 cm, DS/SL)	4.80 gig (5.16 BB)
DVD-RAM 2.0 (12 cm, SS/SL)	4.37 gig (4.70 BB)*
DVD-RAM 2.0 (12 cm, DS/SL)	8.75 gig (9.40 BB)*
DVD-RAM 2.0 (8 cm, SS/SL)	1.36 gig (1.46 BB)*
DVD-RAM 2.0 (8 cm, DS/SL)	2.47 gig (2.65 BB)*
CD-ROM (12 cm, SS/SL)	0.635 gig (0.650 BB)
CD-ROM (8 cm, SS/SL)	0.180 gig (0.194 BB)
DDCD-ROM (12 cm, SS/SL)	1.270 gig (1.364 BB)
DDCD-ROM (8 cm, SS/SL)	0.360 gig (0.387 BB)

*Formatted DVD-RAM discs have slightly less than the stated capacity. For example, the contents of a completely full DVD-R will not quite fit on a DVD-RAM.

TIP: It takes about 2 gigabytes to store 1 hour of average video.

The increase in capacity from CD-ROMs is due to a smaller pit length (~2.08x), tighter tracks (~2.16x), a slightly larger data area (~1.02x), more efficient channel bit modulation (~1.06x), more efficient error correction (~1.32x), and less sector overhead (~1.06x). The total increase for a single layer is about seven times a standard CD-ROM. A slightly different explanation can be found at www.mpeg.org/MPEG/DVD/General/Gain.html.

The capacity of a dual-layer disc is slightly less than double that of a single-layer disc. The laser has to read "through" the outer layer to the inner layer (a distance of 20 to 70 microns). To reduce interlayer crosstalk, the minimum pit length of both layers is increased from 0.4 to 0.44 micrometers. To compensate, the reference scanning velocity is slightly faster, 3.84 meters per second, as opposed to 3.49 meters per second for single-layer discs. Longer pits, spaced farther apart, are easier to read correctly and are less susceptible to jitter. The increased length means fewer pits per revolution, which results in a reduced capacity per layer.

NOTE: Older versions of Windows that use FAT16 instead of UDF, FAT32, or NTFS to read a DVD may run into problems with the 4 gigabyte volume size limit. FAT16 also has a 2 gigabyte file size limit, while FAT32 has a 4 gigabyte file size limit. (NTFS has a 2 TB limit, so we're okay there for a while.)

See "What About Recordable DVDs: DVD-R, DVD-RAM, DVD-RW, DVD+RW, and DVD+R?" in Chapter 4 for details of writable DVDs. More info on disc specifications and manufacturing can be found at Disctronics, Cinram, Panasonic, Technicolor, and other disc replicator sites.

When Did Double-Sided, Dual-Layer Discs (DVD-18) Become Available?

These super-sized discs are used for data but are not commonly used for movies. The first commercial DVD-18 title, *The Stand*, was released in October of 1999. A DVD-18 requires a completely different way of creating two layers. A single-sided, dual-layer disc (DVD-9) is produced by putting one data layer on each substrate and gluing the halves together with transparent adhesive so that the pickup laser can read both layers from one side. But in order to get four layers, each substrate needs to hold two. This requires stamping a second data layer on top of the first, a much more complicated prospect. Only a few replicators can make DVD-18s, and the low yield (the number of usable discs in a batch) makes it more difficult and expensive than making DVD-9s.

What Are the Video Details?

DVD-Video is an application of DVD-ROM, according to the DVD Forum specification (see "Who Invtented DVD Technology and Who Owns It? Who Should Be Contacted for Specifications and Licensing?" in Chapter 6, "Miscellaneous"). DVD-Video is also an application of MPEG-1, MPEG-2,

Dolby Digital, DTS, and other formats. This means the DVD-Video format defines subsets of these standards and formats to be applied in practice to make discs intended for DVD-Video players. DVD-ROM can contain any desired digital information, but DVD-Video is limited to certain data types designed for television reproduction.

A disc has one track (or stream) of MPEG-2 *constant bit rate* (CBR) or *variable bit rate* (VBR) compressed digital video. A restricted version of MPEG-2 *Main Profile at Main Level* (MP@ML) is used. SP@ML is also supported. MPEG-1 CBR and VBR video are also allowed, along with 525/60 (NTSC 29.97 interlaced frames per second [fps]) and 625/50 (PAL/SECAM 25 interlaced frames/sec) video display systems. Coded frame rates of 24 fps progressive from film, 25 fps interlaced from PAL video, and 29.97 fps interlaced from NTSC video are typical.

MPEG-2 progressive_sequence is not allowed, but interlaced sequences can contain progressive pictures and progressive macroblocks. In the case of 24 fps source, the encoder embeds MPEG-2 repeat first field flags into the video stream to make the decoder either perform 2-3 pulldown for 60 Hz NTSC displays (actually 59.94 Hz) or 2-2 pulldown (with a resulting 4 percent speedup) for 50 Hz PAL/SECAM displays. In other words, the player doesn't know what the encoded rate is; it simply follows the MPEG-2 encoder's instructions to produce the predetermined display rate of 25 or 29.97 fps. This is one of the main reasons two kinds of discs are available, one for NTSC and one for PAL. (Very few players convert from PAL to NTSC or NTSC to PAL. Refer to Chapter 1's "Is DVD-Video a Worldwide Standard? Does It Work with NTSC, PAL, and SECAM?")

Because film transfers for NTSC and PAL typically use the same coded picture rate (24 fps) even though PAL resolution is higher, the PAL version takes more space on the disc. The raw increase before encoding is 20 percent (480 to 576), but the final result is closer to 15 percent, depending on encoder efficiency. This translates to an increase of 600 to 700 MB on PAL discs compared to NTSC discs.

It's interesting to note that even interlaced source video is often encoded as progressive-structured MPEG pictures, with interlaced field-encoded macroblocks used only when needed for motion. Most film source is encoded progressive (the inverse telecine process during encoding removes duplicate 2-3 pulldown fields from videotape source), and most video sources are encoded interlaced. These may be mixed on the same disc, such as an interlaced logo followed by a progressive movie.

See "What Is the Difference Between Interlaced and Progressive Video?" for an explanation of progressive and interlaced scanning. See Chapter 1's "What's a Progressive DVD Player?" for progressive-scan players. See the MPEG page at www.mpeg.org for more information on MPEG-2 video.

Picture dimensions are at maximum 720×480 (for 525/60 NTSC display) or 720×576 (for 625/50 PAL/SECAM display). Pictures are subsampled from 4:2:2 ITU-R BT.601 down to 4:2:0 before encoding, allocating an average of 12 bits per pixel in Y'CbCr format. (Color depth is 24 bits, because color samples are shared across 4 pixels.) DVD pixels are not square (see "What's Widescreen? How Do the Aspect Ratios Work?").

The uncompressed source is 124.416 Mbps for video source (720×480×12×30 or 720×576×12×25) or 99.533 or 119.439 Mbps for film source (720×480×12×24 or 720×576×12×24). In analog output terms, lines of horizontal resolution are usually around 500, but they can go up to 540 (see "What Do Lines of Resolution Mean?"). The typical luma frequency response maintains full amplitude between 5.0 and 5.5 MHz. This is below the 6.75 MHz native frequency of the MPEG-2 digital signal (in other words, most players fall short of reproducing the full quality of a DVD). Chroma frequency response is half that of luma.

Allowable picture resolutions are as follows:

MPEG-2, 525/60 (NTSC)   720×480, 704×480, 352×480, 352×240

MPEG-2, 625/50 (PAL)   720×576, 704×576, 352×576, 352×240

MPEG-1, 525/60 (NTSC)   352×240

MPEG-1, 625/50 (PAL)   352×288

Different players use different numbers of bits for the video digital-to-analog converter, with the best-quality players using 10 or 12 bits. This has nothing to do with the MPEG decoding process, because each original component signal is limited to 8 bits per sample. More bits in the player provide more "headroom" and more signal levels during the digital-to-analog conversion, which can help produce a better picture.

The maximum video bit rate is 9.8 Mbps. The average video bit rate is around 4 Mbps but depends entirely on the length, quality, amount of audio, and so on. This is a 31:1 reduction from an uncompressed 124 Mbps video source (or a 25:1 reduction from a 100 Mbps film source). Raw channel data is read off the disc at a constant 26.16 Mbps. After 8/16 demodulation, it's down to 13.08 Mbps. After error correction, the user data stream goes into the track buffer at a constant 11.08 Mbps. The track buffer feeds system stream data out at a variable rate of up to 10.08 Mbps. After system overhead, the maximum rate of combined elementary streams (audio + video + subpicture) is 10.08. MPEG-1 video rate is limited to 1.856 Mbps with a typical rate of 1.15 Mbps.

Still frames (encoded as MPEG I-frames) are supported and can be displayed for a specific amount of time or indefinitely. These are used for menus or slideshows. Still frames can be accompanied by audio.

A disc can also have up to 32 subpicture streams that overlay the video for subtitles, captions for the hard of hearing, captions for children, karaoke, menus, and simple animation. These are full-screen, run-length-encoded bitmaps with 2 bits per pixel, giving 4 color values, and 4 transparency values. For each group of subpictures, 4 colors are selected from a palette of 16 (from the YCbCr gamut), and 4 contrast values are selected out of 16 levels from transparent to opaque. Because one of the four values is usually 100 percent transparency (to let the video show through), only three combinations of colors and transparencies are left, making overlay graphics rather crude. Subpicture display command sequences can be used to create effects such as scrolling, movement, color/highlights, and fades. The maximum subpicture data rate is 3.36 Mbps, with a maximum size per frame of 53,220 bytes.

In addition to subtitles in subpicture streams, DVD also supports NTSC Closed Captions. Closed Caption text is stored in the video stream as MPEG-2 user data (in packet headers) and is regenerated by the player as a line-21 analog waveform in the video signal, which must be decoded by a Closed Caption device in the television. Although the DVD-Video spec mentions NTSC only, no technical reason exists why PAL/SECAM DVD players could not be made to output the Closed Caption text in the *World System Teletext* (WST) format. The only trick is to deal with frame rate differences. An unfortunate note is that DVD Closed Caption MPEG-2 storage format is slightly different than the ATSC format. See Chapter 1's "What's the Difference Between Closed Captions and Subtitles?" for more about Closed Captions.

What Do Lines of Resolution Mean?

Everyone gets confused by the term *lines of horizontal resolution*, also known as LoHR or *TV lines* (TVL). It's a carryover from analog video, is poorly understood, and is inconsistently measured and reported by manufacturers. However, we're stuck with it until all video is digital and we can just report resolution in pixels.

Technically, lines of horizontal resolution refer to *visually resolvable vertical lines per picture height*. In other words, it's measured by counting the number of vertical black and white lines that can be distinguished in an area that is as wide as the picture is high. The idea is to make the measurement independent of the aspect ratio. Lines of horizontal resolution apply both to television displays and to signal formats, such as those produced by a DVD player. Most TVs have ludicrously high numbers listed for their horizontal resolution.

Because DVD has 720 horizontal pixels (on both NTSC and PAL discs), the horizontal resolution can be calculated by dividing 720 by 1.33 (from the

4:3 aspect ratio) to get 540 lines. On a 1.78 (16:9) display, you get 405 lines. In practice, most DVD players provide about 500 lines instead of 540 because of filtering and low-quality digital-to-analog converters. VHS has about 230 (172 widescreen) lines, broadcast TV has about 330 (248 widescreen), and laserdiscs have about 425 (318 widescreen).

Don't confuse lines of horizontal resolution (resolution along the X axis) with scan lines (resolution along the Y axis). DVD produces exactly 480 scan lines of an active picture for NTSC and 576 for PAL. The NTSC standard has 525 total scan lines, but only 480 to 483 or so are visible. (The extra lines contain sync pulses and other information, such as the Closed Captions that are encoded into line 21.) PAL has 625 total scan lines, but only about 576 to 580 are visible. Because all video formats (VHS, laserdisc, and broadcast) have the same number of scan lines, it's the horizontal resolution that makes the big difference in picture quality.

For more information, see Allan Jayne's TV and Video Resolution Explained (http://members.aol.com/ajaynejr/vidres.htm).

What's Widescreen? How Do the Aspect Ratios Work?

Video can be stored on a DVD in 4:3 format (the standard TV shape) or 16:9 (widescreen). The width-to-height ratio of standard television is 4 to 3; in other words, 1.33 times wider than high. New widescreen televisions, specifically those designed for HDTV, have a ratio of 16 to 9, that is, 1.78 times wider than high.

DVD is specially designed to support widescreen displays. Widescreen 16:9 video, such as from a 16:9 video camera, can be stored on the disc in *anamorphic* form, meaning the picture is squeezed horizontally to fit the standard 4:3 rectangle. It is then unsqueezed during playback.

Things get more complicated when film is transferred to video, because most movies today have an aspect ratio of 1.66, 1.85 (flat), or 2.40 (scope). Because these don't match 1.33 or 1.78 TV shapes, two processes are employed to make various movie pegs fit TV holes.

Letterbox (often abbreviated to LBX) means the video is presented in its theatrical aspect ratio, which is wider than standard or widescreen TV. Black bars, called *mattes*, are used to cover the gaps at the top and bottom. A 1.85 movie that has been letterboxed for a 1.33 display has thinner mattes than a 2.4 movie letterboxed to 1.33 (28 percent of the display height versus 44 percent), although the former are about the same thickness as those of a 2.4 movie letterboxed to 1.78 (26 percent of display height). The mattes used to letterbox a 1.85 movie for 1.78 display are so thin (2 percent) that they're hidden by the overscan of most widescreen

TVs. Some movies, especially animated features and European films, have an aspect ratio of 1.66, which can be letterboxed for 1.33 display or *side-boxed* (*windowboxed*) for a 1.78 display.

Pan and scan means the thinner TV "window" is panned across the wider movie picture, chopping off the sides. However, most movies today are shot *soft matte*, which means a full 1.33 aspect film frame is used. (The cinematographer has two sets of frame marks in his or her viewfinder, one for 1.33 and one for 1.85, so he or she can allow for both formats.) The top and bottom are masked off in the theater, but when the film is transferred to video, the full 1.33 frame can be used in the pan and scan process. Pan and scan is primarily used for 1.33 formatting, not for 1.78 formatting, because widescreen fans prefer that letterboxing be used to preserve the theatrical effect. For more details and nice visual aids, see Leopold's How Film Is Transferred to Video page (www.cs.tut.fi/~leopold/Ld/FilmToVideo/). A list of movie aspect ratios is at The Widescreen Movie Center (www.wide-movies.com).

Once the video is formatted to a full-screen or widescreen format, it's encoded and stored on DVD discs. DVD players have four playback modes, one for 4:3 video and three for 16:9 video:

- Full frame (4:3 video for 4:3 display)
- Auto letterbox (16:9 anamorphic video for 4:3 display)
- Auto pan and scan (16:9 anamorphic video for 4:3 display)
- Widescreen (16:9 anamorphic video for 16:9 display)

Video stored in 4:3 format is not changed by the player. It appears normally on a standard 4:3 display. Widescreen systems either enlarge it or add black bars to the sides. 4:3 video may have been formatted with letterboxing or pan and scan before being transferred to DVD. All formatting done to the video prior to it being stored on the disc is transparent to the player. It merely reproduces it as a standard 4:3 TV picture. Video that is letterboxed before being encoded can be flagged so that the player will tell a widescreen TV to automatically expand the picture. Unfortunately, some discs (such as *Fargo*) do not flag the video properly, and worse, some players ignore the flags.

The beauty of anamorphosis is that less of the picture is wasted on letterbox mattes. DVD has a frame size designed for a 1.33 display, so the video still has to be made to fit, but because it's only squeezed horizontally, 33 percent more pixels (25 percent of the total pixels in a video frame) are used to store an active picture instead of black mattes. Anamorphic video is best displayed on widescreen equipment, which stretches the video back to its original width. Alternatively, many new 4:3 TVs can reduce the vertical scan area to restore the proper aspect ratio without losing resolution (an automatic trigger signal is sent to European TVs on SCART pin 8). Even though almost all computers have 4:3 monitors, they have a higher resolution than TVs so they can display the full widescreen picture in a window (854×480 pixels or bigger for NTSC, and 1024×576 or bigger for PAL).

Anamorphic video can be converted by the player for display on standard 4:3 TVs in letterbox or pan and scan form. If anamorphic video is shown unchanged on a standard 4:3 display, people will look tall and skinny as if they have been on a crash diet. DVD players' setup options allow viewers to indicate whether they have a 16:9 or 4:3 TV. In the case of a 4:3 TV, a second option lets the viewer indicate a preference for how the player will reformat anamorphic video: automatic letterbox or automatic pan and scan.

For automatic letterbox mode, the player generates black bars at the top and bottom of the picture (60 lines each for NTSC, 72 for PAL). This leaves three-quarters of the height remaining, creating a shorter but wider rectangle (1.78:1). In order to fit this shorter rectangle, the anamorphic picture is squeezed vertically using a *letterbox filter* that combines every 4 lines into 3, reducing the vertical resolution from 480 scan lines to 360. (If the video was already letterboxed to fit the 1.78 aspect, the mattes generated by the player will extend the mattes in the video.) The vertical squeezing exactly compensates for the original horizontal squeezing so that the movie is shown in its full width. Some players have better letterbox filters than others, using weighted averaging to combine lines (scaling four lines into three or merging the boundary lines) rather than simply dropping one out of every four lines. Widescreen video can be letterboxed to 4:3 on expensive studio equipment before it's stored on the disc, or it can be stored in anamorphic form and letterboxed to 4:3 in the player. If you compare the two, the letterbox mattes will be identical, but the picture quality of the studio version may be slightly better. (See Chapter 1's "How Do I Get Rid of the Black Bars at the Top and Bottom?" for more about letterboxing.)

For automatic pan and scan mode, the anamorphic video is unsqueezed to 16:9, and the sides are cropped off so that a portion of the image is shown at its full height on a 4:3 screen by following a *center of interest* offset encoded in the video stream according to the preferences of those who transferred the film to video. The pan and scan "window" is 75 percent of the full width, which reduces the horizontal pixels from 720 to 540. The pan and scan window can only travel laterally. This does not duplicate a true pan and scan process in which the window can also travel up and down and zoom in and out.

Auto pan and scan has three strikes against it: it doesn't provide the same artistic control as studio pan and scan, a loss of detail occurs when the picture is scaled up, and equipment for recording picture shift information is not widely available. Therefore, no anamorphic movies have been released with auto pan and scan enabled, although a few discs use the pan and scan feature in menus so that the same menu video can be used in both widescreen and 4:3 mode. In order to present a quality full-screen picture to the vast majority of TV viewers yet still provide the best experience for widescreen owners, some DVD producers choose to put two versions on a single disc: 4:3 studio pan and scan and 16:9 anamorphic.

The playback of widescreen material can be restricted by the disc producer. Programs can be marked for the following display modes:

- 4:3 full frame

- 4:3 letterbox (for sending a letterbox expand signal to a widescreen TV)

- 16:9 letterbox only (the player is not allowed to pan and scan on a 4:3 TV)

- 16:9 pan and scan only (the player is not allowed to letterbox on a 4:3 TV)

- 16:9 letterbox or pan and scan (the viewer can select pan and scan or letterbox on a 4:3 TV)

You can usually tell if a disc contains anamorphic video if the package says "enhanced for 16:9 widescreen" or something similar. If all it says is widescreen, it may be letterboxed to 4:3, not 16:9. Widescreen Review has a list of anamorphic DVD titles.

Additional explanations of how anamorphic video works can be found on the Web at Greg Lovern's What's an Anamorphic DVD? page, Bill Hunt's Ultimate Guide to Anamorphic Widescreen DVD, David Lockwood's What Shape Image?, and Dan Ramer's What the Heck Is Anamorphic?. More information can be found at the Anamorphic Widescreen Support Page and the Letterbox/Widescreen Advocacy Page. You might also be interested in Guy Wright's The Widescreen Scam. See Chapter 1 for further discussion of letterboxing.

Anamorphosis causes no problems with line doublers and other video scalers, which simply duplicate the scan lines before they are stretched out by the widescreen display.

For anamorphic video, the pixels are fatter. Different pixel aspect ratios (none of them square) are used for each aspect ratio and resolution; 720-pixel and 704-pixel sizes have the same aspect ratio because the first includes overscan. Note that conventional values of 1.0950 and 0.9157 are for height and width (and are tweaked to match scanning rates). The following minitable uses less confusing width/height values (y/x $\times$ h/w).

	720×480 704×480	720×576 704×576	352×480	352×576
4:3	0.909	1.091	1.818	2.182
16:9	1.212	1.455	2.424	2.909

For the gory details of video resolution and pixel aspect ratios, see Jukka Aho's Quick Guide to Digital Video Resolution and Aspect Ratio Conversions (www.iki.fi./znark/video/conversion).

What Are the Audio Details?

DVD comes in two home-entertainment flavors: DVD-Video and DVD-Audio. Each supports high-definition multichannel audio, but DVD-Audio includes higher-quality PCM audio.

Details of DVD-Audio and *Super Audio* CD (SACD)

Linear Pulse-Code Modulation (LPCM) is mandatory in DVD-Audio discs, with up to six channels at sample rates of 48/96/192 kHz (also 44.1/88.2/176.4 kHz) and sample sizes of 16/20/24 bits. This enables a theoretical frequency response of up to 96 kHz and a dynamic range of up to 144 dB. Multichannel PCM is downmixable by the player, although at 192 and 176.4 kHz only two channels are available. Sampling rates and sizes can vary for different channels by using a predefined set of groups. The maximum data rate is 9.6 Mbps.

The DVD Forum's Working Group 4 (WG4) decided to include lossless compression, and on August 5, 1998 approved Meridian's *Meridian Lossless Packing* (MLP) scheme, licensed by Dolby. MLP removes redundancy from the signal to achieve a compression ratio of about 2:1 while allowing the PCM signal to be completely recreated by the MLP decoder (required in all DVD-Audio players). MLP enables playing times of about 74 to 135 minutes of 6-channel 96 kHz/24-bit audio on a single layer (compared to 45 minutes without packing). Two-channel 192-kHz/24-bit playing times are about 120 to 140 minutes (compared to 67 minutes without packing).

Other audio formats of DVD-Video (Dolby Digital, MPEG audio, and DTS, described later) are optional on DVD-Audio discs, although Dolby Digital is required for audio content that has associated video. A subset of DVD-Video features (no angles and no seamless branching, and so on) is also allowed. Most DVD-Audio players are "universal" players that play DVD-Video discs as well.

DVD-Audio includes specialized downmixing features for PCM channels. Unlike DVD-Video, where the decoder determines how to mix from six channels down to two, DVD-Audio includes coefficent tables to control the mixdown and avoid volume buildup from channel aggregation. Up to 16 tables can be defined by each audio title set (album), and each track can be identified with a table. Coefficients range from 0 dB to 60 dB. This feature goes by the horribly contrived name of *system-managed audio resource technique* (SMART). (Dolby Digital, supported in both DVD-Audio and DVD-Video, also includes downmixing information that can be set at encode time.)

DVD-Audio provides up to 99 still images per track (at typical compression levels, about 20 images fit into the 2 MB buffer in the player), with limited transitions (cut in/out, fade in/out, dissolve, and wipe). Unlike DVD-Video, the user can move at will through the slides without interrupting the audio as it plays; this is called a browsable slideshow. Onscreen displays can be used for synchronized lyrics and navigation menus. A special simplified navigation mode can be used on players without a video display.

Sony and Philips are promoting SACD, a competing DVD-based format using *direct stream digital* (DSD) encoding with sampling rates of 2.8224 MHz. DSD is based on the *pulse-density modulation* (PDM) technique that uses single bits to represent the incremental rise or fall of the audio waveform. This supposedly improves quality by removing the brick wall filters required for PCM encoding. It also makes downsampling more accurate and efficient. DSD provides a frequency response from DC to over 100 kHz with a dynamic range of over 120 dB. DSD includes a lossless encoding technique that produces approximately a 2:1 data reduction by predicting each sample and then run-length encoding the error signal. The maximum data rate is 2.8 Mbps.

SACD includes a physical watermarking feature, *pit signal processing* (PSP), which modulates the width of pits on the disc to store a digital watermark (data is stored in the pit length). The optical pickup must contain additional circuitry to read the PSP watermark, which is compared to information on the disc to make sure it's legitimate. Because of the requirement for new watermarking circuitry, protected SACD discs are not playable in existing DVD-ROM drives.

SACD includes text and still graphics, but no video. Sony says the format is aimed at audiophiles and is not intended to replace the audio CD format. See "What About DVD-Audio or Music DVDs?" in Chapter 1 for more general info on DVD-Audio and SACD.

Audio Details of DVD-Video

The following details are for audio tracks on DVD-Video. Some DVD manufacturers such as Pioneer are developing audio-only players using the DVD-Video format. Some DVD-Video discs contain mostly audio with only still video frames.

A DVD-Video disc can have up to eight audio tracks (streams) associated with each video track (or each video angle). Each audio track can be in one of three formats:

- Dolby Digital (AC-3)   1 to 5.1 channels
- MPEG-2 audio   1 to 5.1 or 7.1 channels
- PCM   1 to 8 channels.

Two additional optional formats are provided: DTS and *Sony Dynamic Digital Sound* (SDDS). Both require the appropriate decoders and are not supported by all players.

The ".1" refers to a *low-frequency effects* (LFE) channel that connects to a subwoofer. This channel carries an emphasized bass audio signal.

Linear PCM is uncompressed (lossless) digital audio, the same format used on CDs and most studio masters. It can be sampled at 48 or 96 kHz with 16, 20, or 24 bits per sample, with 1 to 8 channels used. (Audio CDs are limited to 44.1 kHz at 16 bits.) The maximum bit rate is 6.144 Mbps, which limits sample rates and bit sizes when five or more channels are used.

It's generally felt that the 120 dB dynamic range of 20 bits, combined with a frequency response of around 22,000 Hz from 48 kHz sampling, is adequate for high-fidelity sound reproduction. However, additional bits and higher sampling rates are useful in audiophile applications, studio work, noise shaping, advanced digital processing, and three-dimensional sound field reproduction. DVD players are required to support all the variations of LPCM, but many subsample 96 kHz down to 48 kHz, and some may not use all 20 or 24 bits. The signal provided on the digital output for external digital-to-analog converters may be limited to less than 96 kHz and less than 24 bits.

Dolby Digital is multichannel digital audio, using lossy AC-3 coding technology from original PCM source with a sample rate of 48 kHz at up to 24 bits. The bitrate is 64 kbps to 448 kbps, with 384 or 448 being the normal rate for 5.1 channels, and 192 being the typical rate for stereo (with or without surround encoding). (Most Dolby Digital decoders support up to 640 Kbps, so nonstandard discs with 640 kbps tracks play on many players.) The channel combinations are (front/surround) 1/0, 1+1/0 (dual mono), 2/0, 3/0, 2/1, 3/1, 2/2, and 3/2. The LFE channel is optional with all eight combinations. For details, see the *Advanced Television Systems Committee* (ATSC) document A/52 at www.atsc.org/document.html. Dolby Digital is the format used for audio tracks on almost all DVDs.

MPEG audio is multichannel digital audio, using lossy compression from original PCM format with a sample rate of 48 kHz at 16 or 20 bits. Both

MPEG-1 and MPEG-2 formats are supported. The variable bit rate is 32 to 912 kbps, with 384 being the normal average rate. MPEG-1 is limited to 384 kbps. Channel combinations are (front/surround) 1/0, 2/0, 2/1, 2/2, 3/0, 3/1, 3/2, and 5/2. The LFE channel is optional with all combinations. The 7.1 channel format adds left-center and right-center channels, but it will probably be rare for home use.

MPEG-2 surround channels are in an extension stream matrixed onto the MPEG-1 stereo channels, which makes MPEG-2 audio backwards compatible with MPEG-1 hardware (an MPEG-1 system will only see the two stereo channels.) MPEG Layer 3 (MP3) and MPEG-2 AAC (also known as NBC or unmatrix) are not supported by the DVD-Video standard. MPEG audio is not used much on DVDs, although some inexpensive DVD recording software uses MPEG audio, even on NTSC discs, which goes against the DVD standard and is not supported by all NTSC players.

DTS Digital Surround is an optional multichannel digital audio format, using lossy compression from PCM at 48 kHz at up to 24 bits. The data rate is from 64 to 1536 kbps, with typical rates of 754.5 and 1509.25 for 5.1 channels and 377 or 754 for 2 channels. (The DTS Coherent Acoustics format supports a variable data rate for lossless compression up to 4096 kbps, but this isn't supported on DVDs. DVDs also do not allow DTS sampling rates other than 48 kHz.) Channel combinations are (front/surround) 1/0, 2/0, 3/0, 2/1, 2/2, and 3/2. The LFE channel is optional with all combinations.

DTS ES supports 6.1 channels in two ways: through a Dolby Surround EX-compatible matrixed rear-center channel, or through a discrete seventh channel. DTS also has a 7.1-channel mode (8 discrete channels), but no DVDs have used it yet. The seven-channel and eight-channel modes require a new decoder.

The DVD standard includes an audio stream format reserved for DTS, but many older players ignore it. The DTS format used on DVDs is different from the one used in theaters (Audio Processing Technology's apt-X, which is an ADPCM coder, not a psycho-acoustic coder). All DVD players can play DTS audio CDs, because the standard PCM stream holds the DTS code. See "What's the Deal with DTS and DVDs?" in Chapter 1 for general DTS information. For more info, visit www.dtstech.com and read Adam Barratt's article at home.clearnet.nz/pages/adbarr/page1.html.

SDDS is an optional multichannel (5.1 or 7.1) digital audio format, compressed from PCM at 48 kHz. The data rate can go up to 1280 kbps. SDDS is a theatrical film soundtrack format based on the *Adaptive Transform Acoustic Coding* (ATRAC) compression format also used by Minidisc. Sony has not announced any plans to support SDDS on DVD.

THX (*Tomlinson Holman Experiment*) is not an audio format. It's a certification and quality control program that applies to sound systems and acoustics in theaters, home equipment, and digital mastering processes. The LucasFilm THX Digital Mastering program uses a patented process to track video quality through the multiple video generations needed to make a final format disc or tape, set up video monitors to ensure the filmmaker is seeing a precise rendition of what is on tape before approving the master, and other steps along the way.

THX-certified 4.0 amplifiers enhance Dolby Pro Logic with the following: a crossover that sends bass from front channels to the subwoofer; re-equalization on front channels (to compensate for the high-frequency boost in the theater mix designed for speakers behind the screen); timbre matching on the rear channels; decorrelation of rear channels; a bass curve that emphasizes low frequencies. THX-certified 5.1 amplifiers enhance Dolby Digital and improve on 4.0 with the following: rear speakers are full range, so the crossover sends bass from both front and rear to the subwoofer; decorrelation is turned on automatically when rear channels have the same audio, but not during split-surround effects, which don't need to be decorrelated. More info can be found at the Home THX Program Overview (www.thx.com/consumer_products/home_overview.html).

Discs containing 525/60 (NTSC) video must use PCM or Dolby Digital on at least one track. Discs containing 625/50 (PAL/SECAM) video must use PCM or MPEG audio or Dolby Digital on at least one track. Additional tracks may be in any format. A few first-generation players, such as those made by Matsushita, can't output MPEG-2 audio to external decoders.

The original spec required either MPEG audio or PCM on 625/50 discs. A brief scuffle was led by Philips when early discs came out with only two-channel MPEG and multichannel Dolby Digital, but the DVD Forum clarified in May of 1997 that only stereo MPEG audio was mandatory for 625/50 discs. In December of 1997, the lack of MPEG-2 encoders (and decoders) was a big enough problem that the spec was revised to allow Dolby Digital audio tracks to be used on 625/50 discs without MPEG audio tracks.

Because of the 4 percent speedup from 24 fps film to 25 fps PAL display, the audio must be adjusted to match before it is encoded. Unless the audio is digitally processed to shift the pitch back to normal, it will be slightly high (about half a semitone).

For stereo output (analog or digital), all players have a built-in two-channel Dolby Digital decoder that *downmixes* from 5.1 channels (if present on the disc) to Dolby Surround stereo. That is, five channels are phase matrixed into two channels to be decoded to four channels by a Dolby Pro Logic processor or five channels by a Pro Logic II processor. PAL players also have an MPEG or MPEG-2 audio decoder. Both Dolby Digital and MPEG-2 support two-channel Dolby Surround as the source in cases where the disc producer can't or doesn't want to remix the original onto discrete channels. This means that a DVD labeled as having Dolby Digital sound may only use the left and right channels for surround or plain stereo. Even movies with old monophonic soundtracks may use Dolby Digital, with only one or two channels. Some players can optionally downmix to non-surround stereo. If surround audio is important to you, you will hear significantly better results from multichannel discs if you have a Dolby Digital system.

The new *Dolby Digital Surround EX* (DD-EX) format, which adds a rear center channel, is compatible with DVD discs and players, as well as with existing Dolby Digital decoders. The new *DTS Digital Surround ES* (DTS-ES) format, which likewise adds a rear center channel, works with existing DTS decoders and with DTS-compatible DVD players. However, for full use of either new format, you need a new decoder to extract the rear center channel, which is phase matrixed into the two standard rear channels in the same way Dolby Surround is matrixed into standard stereo channels. Without a new decoder, you'll get the same 5.1-channel audio you get now. Because the additional rear channel isn't a full-bandwidth discrete channel, it's appropriate to call the new formats "5.2-channel" digital surround. There is also DTS-ES Discrete, which adds a full-bandwidth discrete rear center channel in an extension stream that is used by DTS-ES Discrete decoders but ignored by older DTS decoders. DTS-ES decoders include DTS Neo:6, which is not an encoding format but a matrix decoding process that provides 5 or 6 channels.

The Dolby Digital downmix process does not usually include the LFE channel and may compress the dynamic range in order to improve dialog audibility and keep the sound from becoming muddy on average home audio systems. This can result in reduced sound quality on high-end audio systems. The downmix is auditioned when the disc is prepared, and if the result is not acceptable, the audio may be tweaked or a separate left or right Dolby Surround track may be added. Experience has shown that minor tweaking is sometimes required to make the dialog more audible within the limited dynamic range of a home stereo system. Some disc producers include a separately mixed stereo track rather than fiddle with the surround mix.

The Dolby Digital *dynamic range compression* (DRC) feature, often called *midnight mode*, reduces the difference between loud and soft sounds so that you can turn the volume down to avoid disturbing others yet still hear the detail of quiet passages. Some players have the option to turn off DRC.

Dolby Digital also includes a feature called *dialog normalization* (DN), which should more accurately be called volume standardization. DN is designed to keep the sound level the same when switching between different sources. This will become more important as additional Dolby Digital sources (digital satellite and DTV) become common. Each Dolby Digital track contains loudness information so that the receiver can automatically adjust the volume, such as turning it down, for example, during a loud commercial. (Of course, the commercial makers can cheat and set an artificially low DN level, causing your receiver to turn up the volume during the commercial.) Turning DN on or off on your receiver has no effect on the dynamic range or sound quality; its effect is no different than turning the volume control up or down.

All five DVD-Video audio formats support karaoke mode, which has two channels for stereo (left and right) plus an optional guide melody channel (M) and two optional vocal channels (V1 and V2).

A DVD-5 with only one surround stereo audio stream (at 192 kbps) can hold over 55 hours of audio. A DVD-18 can hold over 200 hours.

For more information about multichannel surround sound, see Bobby Owsinski's FAQ at www.surroundassociates.com/fqmain.html.

Can You Explain This Dolby Digital, Dolby Surround, Dolby Pro Logic, and DTS Stuff in Plain English?

Almost every DVD contains audio in the *Dolby Digital* (AC-3) format. *DTS* is an optional audio format that can be added to a disc in addition to Dolby Digital audio. Both DTS and Dolby Digital can store mono, stereo, and multichannel audio (usually 5.1 channels).

Every DVD player in the world has an internal Dolby Digital decoder. The built-in two-channel decoder turns Dolby Digital into stereo audio, which can be fed to almost any type of audio equipment (receiver, TV, boombox, and so on) as a standard analog stereo signal using a pair of stereo audio cables or as a digital PCM audio signal using a coax or optical cable. Refer to "How Do I Hook up a DVD Player?" for more information.

A standard audio mixing technique called *Dolby Surround* "piggybacks" a rear channel and a center channel onto a two-channel signal. A Dolby Surround signal can be played on any stereo system (or even a mono system), in which case the rear- and center-channel sounds remain mixed in with the left and right channels. When a Dolby Surround signal is played on a multichannel audio system that knows how to handle it, the extra channels are extracted to feed center speakers and rear speakers. The original technique of decoding Dolby Surround, called simply *Dolby Surround*, extracts only the rear channel. The improved decoding technique, *Dolby Pro Logic*, also extracts the center channel. A brand-new decoding technology, *Dolby Pro Logic II*, extracts both the center channel and the rear channel and also processes the signals to create more of a 3-D audio environment.

Dolby Surround is independent of the storage or transmission format. In other words, a two-channel Dolby Surround signal can be analog audio, broadcast TV audio, digital PCM audio, Dolby Digital, DTS, MP3, audio on a VHS tape and so on.

Unlike Dolby Surround, Dolby Digital encodes each channel independently. Dolby Digital can carry up to five channels (left, center, right, left surround, right surround) plus an omnidirectional low-frequency channel. The built-in, two-channel Dolby Digital decoder in every DVD player handles multichannel audio by *downmixing* it to two channels using Dolby Surround (refer to "Audio Details of DVD-Video"). This allows the analog stereo outputs to be connected to just about anything, including TVs and receivers with Dolby Pro Logic capabilities. Most DVD players also output the downmixed two-channel Dolby Surround signal in digital PCM format, which can be connected to a digital audio receiver, most of which do Dolby Pro Logic decoding.

Most DVD players also output the raw Dolby Digital signal for connection to a receiver with a built-in Dolby Digital decoder. Some DVD players have built-in multichannel decoders to provide six (or seven) analog audio outputs to feed a receiver or amplifier with multichannel analog inputs. Refer to "What Are the Outputs of a DVD Player?" for more info.

DTS is handled differently. Many DVD players have a *DTS Digital Out* feature (also called DTS passthrough) that sends the raw DTS signal to an external receiver with a DTS decoder. A few players have a built-in two-channel DTS decoder that downmixes to Dolby Surround, just like a two-channel Dolby Digital decoder. Some players have a built-in multichannel DTS decoder with six (or seven) analog outputs. Some DVD players don't recognize DTS tracks at all (refer to "What's the Deal with DTS and DVD" in Chapter 1).

If you have a *plain old stereo* (POS), a Dolby Surround receiver, or a Dolby Pro Logic receiver, you don't need anything special in the DVD player. Any model will connect to your system. If you have a Dolby Digital receiver, you need a player with Dolby Digital out (all but the cheapest players have this). If your receiver can also do DTS, you should get a player with DTS Digital Out. The only reason to get a player with six-channel Dolby Digital or DTS decoder output is if you want use multichannel analog connections to the receiver (see the component analog section of "How Do I Hook up a DVD Player?").

Why Is the Audio Level from my DVD Player So Low?

Many people complain that the audio level from DVD players is too low. In truth, the audio level is too high on everything else. Movie soundtracks are extremely dynamic, ranging from near silence to intense explosions. In order to support an increased dynamic range and hit peaks (near the 2-volt RMS limit) without distortion, the average sound volume must be lower.

This is why the line level from DVD players is lower than from almost all other sources. So far, unlike on CDs and laserdiscs, the level is much more consistent between discs. If the change in volume when switching between DVDs and other audio sources is annoying, you can adjust the output signal level on some players, or the input signal level on some receivers, but other than that, you don't have many other options.

How Do the Interactive Features Work?

DVD-Video players (and software DVD-Video navigators) support a command set that provides rudimentary interactivity. The main feature is menus, which are present on almost all discs to allow content selection and feature control. Each menu has a still or motion background and up to 36 highlightable, rectangular buttons (only 12 if widescreen, letterbox, and pan and scan modes are used). Remote control units have up/down and left/right arrow keys for selecting onscreen buttons, along with numeric keys, a select key, a menu key, a top menu (title) key, and a return key. Additional remote functions may include freeze, step, slow, fast, scan, next, previous, audio select, subtitle select, camera angle select, play mode select, search to program, search to part of title (chapter), search to time, and search to camera angle. Any of these features can be disabled by the producer of the disc, an act which is called *user operation control* (UOP). It's commonly used to lock you into the copyright warning or movie previews at the beginning of the disc, or to keep you from changing audio or subtitle tracks during the movie.

Additional features of the command set include simple math (add, subtract, multiply, divide, modulo, and random); bitwise and; bitwise or; bitwise xor; plus comparisons (equal, greater than, and so on); and register loading, moving, and swapping. Twenty-four system registers exist for information such as the language code, audio and subpicture settings, and the parental level. Sixteen general registers are available for command use. A countdown timer is also provided. Commands can branch or jump to other commands. Commands can also control player settings; jump to different parts of the disc; and control the presentation of audio, video, subpicture, camera angles, and so on. The command set enables the creation of relatively sophisticated discs, such as games or interactive educational programs.

DVD-V content is broken into *titles* (movies or albums) and *parts of titles* (chapters or songs). Titles are made up of *cells* grouped into *programs* and are linked together by one or more *program chains* (PGC). A PGC can be one of three types: sequential play, random play (may repeat), or shuffle play (random order but no repeats). Individual cells may be used by more than one PGC, which is how parental management and seamless branching are accomplished: Different PGCs define different sequences through mostly the same material.

Additional material for camera angles and seamless branching is interleaved together in small chunks. The player jumps from chunk to chunk, skipping over unused angles or branches, to stitch together the seamless video. Because angles are stored separately, they have no direct effect on the bit rate, but they do affect the playing time. Adding one camera angle for a program roughly doubles the amount of space needed (and cuts the playing time in half). Examples of branching (seamless and nonseamless) include *Kalifornia*, *Dark Star*, *Stargate Special Edition*, and *The Abyss*.

What Is the Difference Between Interlaced and Progressive Video?

Basically, video can be displayed using two different methods: *interlaced scan* or *progressive scan*. Progressive scan, used in computer monitors and digital TVs, displays all the horizontal lines of a picture at one time as a single *frame*. Interlaced scan, used in standard television formats (NTSC, PAL, and SECAM), displays only half of the horizontal lines at a time (the first *field*, containing the odd-numbered lines, is displayed, followed by the second field, containing the even-numbered lines). Interlacing relies on the phosphor persistence of the TV tube to blend the fields together over a fraction of a second into a seemingly single picture. The advantage of interlaced video is that a high refresh rate (50 or 60 Hz) can be achieved with only half the bandwidth. The disadvantage is that the horizontal resolution is essentially cut in half, and the video is often filtered to avoid flicker (interfield twitter) and other artifacts.

It may help to understand the difference by considering how the source images are captured. A film camera captures full frames in intervals that are 1/24th of a second long, whereas a video camera alternately scans fields of odd and even lines in 1/60th of a second intervals, resulting in interlaced frames that are 1/30th of a second long. (Unlike projected film, which shows the entire frame in an instant, many progressive-scan displays trace a series of lines from top to bottom, but the end result is about the same.)

DVDs are specifically designed to be displayed on interlaced-scan displays, which cover 99.9 percent of the more than one billion TVs worldwide. However, most DVD content comes from film, which is inherently progressive. To make film content work in an interlaced form, the video from each film frame is split into two video fields, 240 lines in one field, and 240 lines in the other, and encoded as separate fields in the MPEG-2 stream.

A complication is that film runs at 24 frames per second, whereas TV runs at 30 frames (60 fields) per second for NTSC, or 25 frames (50 fields) per second for PAL and SECAM. For a PAL/SECAM display, the simple solution is to show the film frames at 25 fps, which is a 4 percent speed increase, and to speed up the audio to match. For an NTSC display, the solution is to spread 24 frames across 60 fields by alternating the display of the first film frame for two video fields and the next film frame for three video fields. This is called *2-3 pulldown.* The sequence works as shown here, where A through D represent film frames; A1, A2, B1, and so on represent the separation of each film frame into two video fields; and 1 through 5 represent the final video frames:

Film frames: | A | B | C | D |

Video fields: |A1 A2 | B1 B2 | B1   C2 | C1 D2 | D1 D2|

Video frames: | 1 | 2 | 3 | 4 | 5 |

For MPEG-2 encoding, repeated fields (B1 and D2) are not actually stored twice. Instead, a flag is set to tell the decoder to repeat the field. (The inverted order of C2 and C1 and D2 and D1 is because of the requirement that top and bottom fields alternate. Because the fields are from the same film frame, the order doesn't matter.) MPEG-2 also has a flag to indicate when a frame is progressive (the two fields come from the same instant in time). For film content, the progressive_frame flag should be true for every frame. (Refer to "What Are the Video Details?" for more MPEG-2 details.)

As you can see, a couple of problems are inherent in 2-3 pulldown. Some film frames are shown for a longer period of time than others, causing *judder*, or jerkiness, that shows up especially in smooth pans. Also, if you freeze the video on the third or fourth video frame when motion is taking place in the picture, you will see two separate images combine to be a flickering mess. Most DVD players avoid the second problem by only pausing on coherent frames or by only showing one field, although some enable you to freeze on flicker frames. (This is what the frame/field still option in the player's setup menu refers to.)

Most DVD players are hooked up to interlaced TVs, so not much can be done about artifacts from film conversion. However, see "What's a Progressive DVD Player?" in Chapter 1 for information about progressive DVD players.

For more on progressive video and DVDs, see Part 5 and Player Ratings in the excellent DVD Benchmark series at "Secrets of Home Theater and High Fidelity" at www.hometheaterhifi.com.

NOTE: 2-3 pulldown is the same term as 3:2 pulldown, but this book uses the 2-3 notation to indicate that it's a sequence, not a ratio, and that in practice two fields are usually created from the first film frame.

What Is Edge Enhancement?

When films are transferred to video in preparation for DVD encoding, they are commonly run through digital processes that attempt to clean up the picture. These processes include *digital video noise reduction* (DVNR) and image enhancement. Enhancement increases the contrast (similar to the effect of the sharpen or unsharp mask filters in Photoshop), but it can tend to overdo areas of transition between light and dark or different colors, causing a chiseled look or a *ringing* effect, such as the haloes you see around streetlights when driving in the rain.

DVNR is a good thing, when done well, because it can remove scratches, spots, and other defects from the original film. Enhancement, which is rarely done well, is a bad thing. The video may look sharper and clearer to the casual observer, but fine tonal details of the original picture are altered and lost.

Note that ringing can also be caused by the player and by the TV. *Scan velocity modulation* (SVM), for example, causes ringing.

Does DVD Work with Barcodes?

If your humble book author and other long-time developers of laserdiscs had prevailed, all DVD players would support barcodes. This would have made for really cool printed supplements and educational material that could jump to any part of a disc with a swipe of a barcode wand. But the rejection of our recommendations after an all-star meeting in August of 1995 is another story for another day.

So the answer is "mostly no." A few industrial players, the Pioneer LD-V7200, Pioneer LD-V7400, and Philips ProDVD-170 support barcodes, including compatibility with the LaserBarCode standard. The DVD must be authored with one sequential PGC titles in order for the timecode search to work. More info can be found in the Pioneer technical manuals (www. pioneeraus.com.au/multimedia/manuals/op_manuals_index.htm).

What Is BCA?

BCA stands for *burst cutting area*, a zone near the hub of a DVD reserved for a barcode that can be etched into the disc by a YAG laser. Because barcode cutting is independent of the stamping process, each disc can have unique data recorded in the BCA, such as a serialized ID. DVD readers can use the laser pickup head to read the BCA. The BCA is used by CPRM (refer to "What Are the Copy Protection Issues?" in Chapter 1) and Divx (see "What is Divx?" in Chapter 2) to uniquely identify each disc.

How Long Do DVDs Last?

Pressed discs (the kind that movies come on) last longer than you will, anywhere from 50 to 300 years. The expected longevity of dye-based DVD-R and DVD+R discs is anywhere from 40 to 250 years, about as long as CD-R discs. The phase-change erasable formats (DVD-RAM, DVD-RW, and DVD+RW) have an expected lifetime of 25 to 100 years.

A good discussion of CD-R longevity and test info can be found at the Kodak web site. Also see www.ee.washington.edu/conselec/CE/kuhn/otherformats/95x9.htm and www.cd-info.com/CDIC/Technology/CD-R/Media/Kodak.html for more info.

For comparison, magnetic media (tapes and disks) last 10 to 30 years, high-quality, acid-neutral paper can last 100 years or longer, and archival-quality microfilm is projected to last 300 years or more. Note that computer storage media often becomes technically obsolete within 20 to 30 years, long before it physically deteriorates. In other words, before the media becomes unviable, it may become difficult or impossible to find equipment that can read it.

What About the HD-DVD and Blue Laser Formats?

HD-DVD (HD stands for both *high-density* and *high-definition*) was under development before DVDs came out and finally emerged in 2003 (refer to "Will High-Definition DVDs or 720p DVDs Make Current Players and Discs Obsolete?" in Chapter 2 for general info). Some high-definition versions of HD-DVD use the original DVD physical format but depend on new video-encoding technology, such as H.264, to fit high-definition video in the space that used to hold only standard-definition video. High-density formats use blue or violet lasers to read smaller pits, increasing the data capacity to around 15 to 30 GB per layer. High-density formats use high-definition MPEG-2 video (for compatibility with ATSC and DVB HD broadcasts, refer to "Do DVDs Support HDTV (DTV)? Will HDTV Make DVDs Obsolete?" in

TABLE 3-2   HD-DVD Proposals

Format	Data depth	Laser	Video	Capacity (single layer/dual layer)	Data rate
HD-DVD-9	0.6 mm	Red (650 nm)	New codec	NA/8.5G (ROM)	11 Mbps
AOD	0.6 mm	Blue (405 nm)	HD MPEG-2 and new codec	15G/30G (ROM), 20G/40G (recordable)	36 Mbps
Blue-HD-DVD-1	0.6 mm	Blue (405 nm)	AVC	17G/NA	25.05 Mbps
Blu-ray	0.1 mm	Blue (405 nm)	HD MPEG-2	27G/50G	36 Mbps
Blue-HD-DVD-2	0.1 mm	Blue (405 nm)	AVC	17G/NA	31.59 Mbps

Chapter 2) and may also use advanced encoding formats, probably supporting 1080p24 video.

As of early 2003, five proposals have been made for HD-DVD, with the possibility of others. Table 3-2 provides a summary.

HD discs will not play on existing players. Even HD-DVD-9 discs, which the player can physically read, require new circuitry to decode and display the HD video. HD-DVD-9 discs could play on DVD PCs with the right software upgrades. Blue-laser discs require new optical assemblies and controllers. HD players will undoubtedly read existing DVDs, so your collection will not become obsolete when you buy a new player. None of the HD formats will be used for movies until 2005 or 2006.

HD-DVD-9, aka HD-9

HD video on existing dual-layer DVD-9 discs, or HD-DVD-9, will require new players to handle the new video encoding format and the higher data rate. The format is under development within the DVD Forum, primarily backed by Warner. It was originally positioned as a transition format to future HD-DVD, but it is now touted as a compatible but cheaper to replicate companion to blue-laser HD-DVD.

A two-hour movie can fit on a DVD-9 at data rates of 6 to 7 Mbps. Given advances in video compression technology, it should be possible to get HD quality of at least 720p24 at these data rates (720 lines of progressive video at 24 fps). Shorter movies could be encoded in 1080p24 format. H.264 (MPEG-4 part 10) is the likely encoding standard.

Advanced Optical Disc (AOD), aka DVD2

The DVD Forum's next-generation DVD is *Advanced Optical Disc* (AOD), or DVD2. AOD is a modification of the existing DVD physical format to enable about 15 GB per layer using a blue-ultraviolet readout laser. The same 0.6-millimeter data depth is used. AOD is designed to improve data capacity while theoretically being able to use existing replication equipment. It is primarily supported by Toshiba and NEC.

Blu-ray Disc (BD)

The *Blu-ray Disc* (BD) is a new high-density physical format that will hold 23 to 27 GB per layer using a blue-ultraviolet laser and a 0.1-millimeter data depth. Because of the 0.1-millimeter cover layer, it requires significant changes to production equipment. Blu-ray is initially intended for home recording, professional recording, and data recording. Mass-market distribution of prerecorded movies will come later, after a read-only format is developed and the details of video, audio, interactivity, and copy protection are hammered out. The Blu-ray backers include LG, Panasonic, Philips, Pioneer, Hitachi, Mitsubishi, Samsung, Sharp, Sony, and Thomson. Sony released the first BD recorder in Japan in April of 2003.

As far as technical details, a BD holds up to 27 GB per layer using a 0.1-millimeter recording depth (to reduce aberration from disc tilt). It uses a 405-nanometer blue-violet semiconductor with 0.85 numerical aperture (NA) lens design to provide a 0.32 μm track pitch (half that of DVDs) and as small as a 0.138 μm pit length. Variations include 23.3 GB of capacity with a 0.160 μm minimum pit length or 25 GB of capacity with a 0.149 μm minimum pit length. The physical discs will use phase-change groove recording on a 12-centimeter diameter, 1.2-millimeter-thick disc, similar to DVD-RW and DVD+RW. It has a 36 Mbps data transfer rate. The recording capacity on a single layer is about 2 hours of HD video (at 28 Mbps) or about 10 hours of standard-definition video (at 4.5 Mbps). The cartridge size is 129 $\times$ 131 $\times$ 7 millimeters. Plans are to produce dual-layer recordable discs, holding about 50 GB per side, but such discs will take a few additional years to appear.

Blue-HD-DVD-1 and Blue-HD-DVD-2

The *Advanced Optical Storage Research Alliance* (AOSRA), formed by Taiwan's *Industrial Technology Research Institute* (ITRI) has its own variations of blue-laser formats. Blue-HD-DVD-1 uses a 0.6-millimeter data depth similar to AOD, and Blue-HD-DVD-2 uses a 0.1-millimeter data depth similar to Blu-ray.

Chapter 4

DVD and Computers

Can I Play DVD Movies on My Computer?

Yes, if your computer has the right stuff. Almost all Windows and Mac OS computers with DVD drives come with software to play DVDs.

The computer operating system or playback software must support regional codes and be licensed to descramble copy-protected movies. If the computer has TV video out, it must support Macrovision in order to play copy-protected movies. You may also need software that can read the UDF file system format used by DVDs. You don't need special drivers for Windows or Mac OS, since the existing CD-ROM drivers work fine with DVD-ROM drives. In addition to a DVD-ROM drive you must have software (or extra hardware) that knows how to play the DVD-Video format and decode MPEG-2 video and Dolby Digital or MPEG-2 audio. Good-quality software-only playback requires a 350-MHz Pentium II or a Mac G4. Almost all new computers with DVD-ROM drives use software decoding instead of hardware decoding. Hardware upgrade kits can be purchased for older computers (usually minimum 133 MHz Pentium or G3), starting at $150.

Mac OS X 10.0 (Cheetah) had no support for DVD playback when released in March 2001, and also did not support Apple's DVD authoring applications (iDVD and DVD Studio Pro). (More info at CNET.) Support for DVD playback was added to version 10.1 (Puma).

If you're having problems playing movies on your computer, see the section, "Why Do I Have Problems Playing DVDs on My Computer?"

Certain MPEG decoding tasks such as motion compensation, IDCT (inverse discrete cosine transform), IVLC (inverse variable length coding), and even subpicture decoding can be performed by special circuitry on a video graphics chip, improving the performance of software decoders. This is called *hardware decode acceleration*, *hardware motion comp*, or *hardware assist*. Some card makers also call it hardware decode, even though they don't do all the decoding in hardware. All modern graphics cards also provide hardware colorspace conversion (YCbCr to RGB) and videoport overlay (some graphics card makers make a big deal about this even though all their competitors' cards have the same feature).

Microsoft Windows 98, 2000, Me, and XP include DirectShow, which provides standardized support for DVD-Video and MPEG-2 playback. DirectShow can also be installed in Windows 95 (it's available for download). DirectShow creates a framework for DVD applications, but a third-party hardware or software decoder is required (see below). Windows NT 4.0 supports DVD-ROM drives for data, but has very little support for playing DVD-Video discs. Margi DVD-To-Go, Sigma Designs Hollywood Plus, and the related Creative Labs Dxr3 are among the few hardware decoders that work in NT 4.0. InterVideo WinDVD software works in NT 4.0 (National Semiconductor DVD Express and MGI SoftDVD Max also work in NT 4.0, but they aren't available retail.) Windows 98 and newer can read UDF discs. Version 6.1 of Windows Media Player enabled scriptable DVD playback in an HTML page (see "How Do I Play DVD Video in HTML, PowerPoint, Director, VB, and So On?" for more on DVD playback control). Version 7 of Windows Media Player dropped all DVD support. Version 8 of Windows Media Player added a user interface for DVD playback, but no scripting. Adaptec provides a free filesystem driver, *UDF Reader*, for Windows 95/98/NT. Software Architects sells *Read DVD* for Windows 95.

Apple QuickTime 6 is partially ready for DVD-Video and MPEG-2 but does not yet have full decoding or DVD-Video playback support in place. Mac OS 8.1 or newer can read UDF discs. Roxio provides a free utility, *UDF Volume Access*, that enables Mac OS 7.6 and newer to read UDF discs. Software Architects sells UDF reading software for Mac OS called *DVD-RAM TuneUp*. Intech's CD/DVD SpeedTools software allows most any DVD drive to be used with a Mac.

NOTE: The QuickTime MPEG Extension for Mac OS is for MPEG-1 only and does not play MPEG-2 DVD-Video.

DVD player applications (using either software or hardware decoding) are virtual DVD players. They support DVD-Video features (menus, subpictures, and so on) and emulate the functionality of a DVD-Video player remote control. Many player applications include additional features such as bookmarks, chapter lists, and subtitle language lists.

Microsoft Windows includes a DVD software player, but does not include the necessary decoder. You must have a third-party software or hardware decoder in order to play a DVD. Most PCs that come with a DVD drive include a decoder, or you can purchase one. See "How Do I Get the Microsoft Windows DVD Player Application to Run?" and "I Upgraded to Windows XP. Why Did My DVD Software Stop Working?" for more info.

Software decoders and DVD player applications for Microsoft Windows PCs:

- ATI: A special version of CineMaster software for certain ATI graphics cards

- ASUS: *ASUSDVD* (a custom version of InterVideo WinDVD software or CyberLink PowerDVD software)

- KiSS: *CoolDVD* (DirectShow [Windows 98/Me/2000/XP])

- Creative Technology: *SoftPC-DVD*

- CyberLink: *PowerDVD* (DirectShow [Windows 98/Me/2000/XP]; NT 4.0; available for purchase)

- ELSA: ELSAMovie, German only

- InterVideo: *WinDVD* (DirectShow [Windows 98/Me/2000/XP]; NT 4.0; available for purchase)

- Matrox: special version of CineMaster software for certain Matrox graphics cards

- National Semiconductor: *DVD Express* (DirectShow [Windows 98/Me/2000/XP]; OEM only)

- NEC (NEC PCs only)

- Odyssey: *Odyssey DVD Player* (available for purchase)

- Orion Studios: *DirectDVD* (DirectShow, downloadable shareware)

- Sonic (formerly Ravisent, formerly Quadrant International): *CinePlayer* (DirectShow [Windows 98/Me/2000/XP]; available for purchase)

- Varo Vision: *VaroDVD*

Xing DVDPlayer is no longer available since the company was purchased by Real Networks

Software decoders need at least a 350 MHz Pentium II and a DVD-ROM drive with bus mastering DMA to play without dropped frames. Anything slower than a 400 MHz Pentium III will benefit quite a bit from hardware decode acceleration in the graphics card. An AGP graphics card (rather than PCI) also improves the performance of software decoders.

Hardware decoder cards and DVD-ROM upgrade kits for Microsoft Windows PCs:

- Creative Technology: *PC-DVD Encore Dxr3*, Sigma EM8300 chip (no DirectShow yet); *PC-DVD Encore Dxr2*, C-Cube chip (DirectShow, Win2000)

- Digital Connection: *3DFusion*, Mpact2 chip (DirectShow)

- Digital Voodoo: *D1 Desktop 64*, Digital Voodoo chip (professional, QuickTime)

- E4 (Elecede): *Cool DVD*, C-Cube chip (E4 has gone out of business)

- IBM: *ThinkPad* laptops, IBM chip (DirectShow)

- LeadTek: *WinFast 3D S800*, Mpact2 chip (DirectShow)

- Margi: *DVD-to-Go*, ZV PC card for laptops (DirectShow, Win2000)

- Ravisent: *Hardware Cinemaster*, C-Cube chip (DirectShow)

- Philips Electronics: *PCDV632, PCVD104* (*K* series come with Sigma *Hollywood* card, *R* series come with software decoder) (DirectShow)

- Samsung: *Revolution*, Samsung SD 606 6x, Sigma *Hollywood Plus* card (DirectShow)

- Sigma Designs: *Hollywood* series, Sigma EM8300 chip (no Direct-Show yet)

- STB: *DVD Theater*, Mpact2 chip (DirectShow)

- Stradis: *Stradis Professional MPEG-2 Decoder*, IBM chip (professional, no DirectShow)

- Toshiba: *Tecra* laptops, C-Cube chip (DirectShow)

- Vela Research: *CineView Pro* (professional, no DirectShow)

All but the Sigma Designs decoder (including Creative Dxr3) have WDM drivers for DirectShow. The Sigma Designs decoder card is used in hardware upgrade kits from Hitachi, HiVal, Panasonic, Phillips, Sony, Toshiba, and VideoLogic. The advantage of hardware decoders is that they don't eat up CPU processing power, and they often produce better quality video than software decoders. The Chromatic Mpact2 chip does 3-field analysis to produce exceptional progressive-scan video from DVDs (unfortunately, Chromatic was bought by ATI and the chip is no longer supported, although some of the technology is now in ATI's Radeon). Hardware decoders use *video overlay* to insert the video into the computer display. Some use analog overlay, which takes the analog VGA signal output from the graphics card and keys in the video, while others use video port extension (VPE), a direct digital connection to the graphics adapter via a cable inside the computer. Analog overlay may degrade the quality of the VGA signal. See "Why Can't I Take a Screenshot of DVD Video? Why Do I Get a Pink or Black Square?" for more overlay info.

Many Macintosh models come standard with DVD-ROM, DVD-RAM, or DVD-RW drives. The included Apple software DVD player uses hardware acceleration in the ATI graphics card. The still-unreleased QuickTime MPEG-2 decoder may use the Velocity Engine (AltiVec) portion of the PowerPC (G4) chip for video and audio decoding. DVD-ROM upgrade kits and decoder cards for Macintoshes were made by E4 (Elecede) (*Cool DVD*, C-Cube chip) [E4 has gone out of business], EZQuest (*BOA Mac DVD*), Fantom Drives (*DVD Home Theater* kit: DVD-ROM or DVD-RAM drive with Wired MPEG-2 card), and Wired (*Wired 4DVD*, Sigma EM8300 chip [same card as Hollywood plus]; *MasonX* [can't play encrypted movies]; *DVD-To-Go* [out of production]; Wired was acquired by Media100 but later reconstituted).

The Sigma Designs *NetStream 2000* DVD decoder card supports Linux DVD playback. InterVideo and CyberLink have also announced DVD player applications for Linux, although the CyberLink player is only available to OEMs. In addition, there are free software players for Linux, Unix, BeOS, and other operating systems: MPlayer, OMS (LiViD), VideoLan, and Xine.

Computers have the potential to produce better video than set-top DVD-Video players by using progressive display and higher scan rates, but many PC systems don't look as good as a home player hooked up to a quality TV.

If you want to hook a DVD computer to a TV, the decoder card or the VGA card must have a TV output (composite video or s-video). Video quality is much better with s-video. Alternatively, you can connect a scan converter to the VGA output. Scan converters are available from ADS Technologies, AITech, Antec, AverLogic, AVerMedia, Communications Specialties, Digital Vision, Focus Enhancements, Key Digital Systems, RGB Products, and others. Make sure the scan converter can handle the display resolution you have chosen: 640x480, 800x600, and so on, although keep in mind that even 800x600 is beyond the ability of a standard TV, so higher resolutions won't make the TV picture better.

The quality of video from a PC depends on the decoder, the graphics card, the TV encoder chip, and other factors. The RGB output of the VGA card in computers is at a different frequency than standard component RGB video, so it can't be directly connected to most RGB video monitors. If the decoder card or the sound card has Dolby Digital or DTS output, you can connect to your A/V receiver to get multichannel audio.

A DVD PC connected to a progressive-scan monitor or video projector, instead of a standard TV, usually looks much better than a consumer player. See "Does DVD Support HDTV (DTV)? Will HDTV Make DVD Obsolete?" in Chapter 2, "DVD's Relationship to Other Products and Technologies." Also see the Home Theater Computers forum at AVS (www.avsforum.com).

For remote control of DVD playback on your PC, check out Animax *Anir Multimedia Magic*, Evation IRMan, InterAct *WebRemote*, Multimedia Studio Miro *MediaRemote*, Packard Bell *RemoteMedia*, RealMagic *Remote Control*, and X10 *MouseRemote*. Many remotes are supported by Visual Domain's *Remote Selector* software.

Can I Play DVD-Audio Discs on My Computer?

Usually not. DVD-ROM drives can read DVD-Audio discs, but as of mid-2003 only the Sound Blaster Audigy 2 card includes the software needed to play DVD-Audio on a computer. Part of the reason for general lack of support is that very few computers provide the high quality audio environment needed to take advantage of DVD-Audio fidelity.

It's possible that Microsoft could add DVD-Audio playback to a future version of Windows, in which case you would only need to download some inexpensive decoding software to get DVD-Audio playback.

What Are the Features and Speeds of DVD-ROM Drives?

Unlike CD-ROM drives, which took years to move up to 2x, 3x, and faster spin rates, faster DVD-ROM drives began appearing in the first year. A 1x DVD-ROM drive provides a data transfer rate of 1.321 MB/s (11.08×10^6/$8/2^{20}$) with burst transfer rates of up to 12 MB/s or higher. The data transfer rate from a DVD-ROM disc at 1x speed is roughly equivalent to a 9x CD-ROM drive (1x CD-ROM data transfer rate is 150 KB/s, or 0.146 MB/s).

DVD physical spin rate is about 3 times faster than CD (that is, 1x DVD spin $\approx$ 3x CD spin), but most DVD-ROM drives increase motor speed when reading CD-ROMs, achieving 12x or faster performance. A drive listed as "16x/40x" reads a DVD at 16 times normal, or a CD at 40 times normal. DVD-ROM drives are available in 1x, 2x, 4x, 4.8x, 5x, 6x, 8x, 10x, and 16x speeds, although they usually don't achieve sustained transfer at their full rating. The "max" in DVD and CD speed ratings means that the listed speed only applies when reading data at the outer edge of the disc, which moves faster. The average data rate is lower than the maximum rate. Most 1x DVD-ROM drives have a seek time of 85-200 ms and access time of 90-250 ms. Newer drives have seek times as low as 45 ms (see Table 4-1).

TABLE 4-1   DVD and CD speed ratings

DVD drive speed	Data rate	Equivalent CD rate	CD reading speed
1x	11.08 Mbps (1.32 MB/s)	9x	8x–18x
2x	22.16 Mbps (2.64 MB/s)	18x	20x–24x
4x	44.32 Mbps (5.28 MB/s)	36x	24x–32x
5x	55.40 Mbps (6.60 MB/s)	45x	24x–32x
6x	66.48 Mbps (7.93 MB/s)	54x	24x–32x
8x	88.64 Mbps (10.57 MB/s)	72x	32x–40x
10x	110.80 Mbps (13.21 MB/s)	90x	32x–40x
16x	177.28 Mbps (21.13 MB/s)	144x	32x–40x

The bigger the cache (memory buffer) in a DVD-ROM drive, the faster it can supply data to the computer. This is useful primarily for data, not video. It may reduce or eliminate the pause during layer changes, but has no effect on video quality.

Rewritable DVD drives (see "What About Recordable DVD: DVD-R, DVD-RAM, DVD-RW, DVD+RW, and DVD+R?") write at about half their advertised speed when the data verification feature is turned on, which reads each block of data after it is written. Verification is usually on by default in DVD-RAM drives. Turning it off will speed up writing. Whether this endangers your data is a subject of debate. Verification is off in DVD-RW and DVD+RW drives.

In order to maintain constant linear density, typical CD-ROM and DVD-ROM drives spin the disc more slowly when reading near the outside where there is more physical surface in each track. (This is called CLV, *constant linear velocity*.) Some faster drives keep the rotational speed constant and use a buffer to deal with the differences in data readout speed. (This is called CAV, *constant angular velocity*.) In CAV drives, the data is read fastest at the outside of the disc, which is why specifications often list "max speed."

NOTE: When playing movies, a fast DVD drive gains you nothing more than possibly smoother scanning and faster searching. Speeds above 1x do not improve video quality from DVD-Video discs. Higher speeds only make a difference when reading computer data, such as when playing a multimedia game or when using a database.

Connectivity of DVD drives is similar to that of CD drives: EIDE (ATAPI), SCSI-2, USB, etc. All DVD drives have analog audio connections for playing audio CDs. No DVD drives have been announced with their own DVD audio or video outputs (which would require internal audio/video decoding hardware).

Almost all DVD-Video and DVD-ROM discs use the *UDF bridge* format, which is a combination of the DVD *MicroUDF* (subset of UDF 1.02) and ISO 9660 file systems. The OSTA UDF file system will eventually replace the ISO 9660 system originally designed for CD-ROMs, but the bridge format provides backwards compatibility until more operating systems support UDF.

What Is the Audio Output Connector on a DVD Drive For?

DVD-ROM drives and DVD recordable drives have an RCA connector or a 4-pin flat (Molex) connector to send analog audio to the audio card in the PC. This is just like the connecter on a CD drive, and in fact it's only for playing audio CDs. The audio from DVDs comes through the computer, not out of the drive. Playing audio from a CD used to require the analog audio output, but most PCs can now play digital audio directly from the CD so the analog connector is not needed.

What About Recordable DVD: DVD-R, DVD-RAM, DVD-RW, DVD+RW, and DVD+R?

There are six recordable versions of DVD-ROM: DVD-R for General, DVD-R for Authoring, DVD-RAM, DVD-RW, DVD+RW, and DVD+R. DVD-R and DVD+R can record data once, similar to CD-R, while DVD-RAM, DVD-RW, and DVD+RW can be rewritten thousands of times, similar to CD-RW. DVD-R was first available in fall 1997. DVD-RAM followed in summer 1998. DVD-RW came out in Japan in December 1999, but was not available in the U.S. until spring 2001. DVD+RW became available in fall 2001. DVD+R was released in mid-2002.

Recordable DVD was first available for use on computers only. Home DVD video recorders (see "Can DVD record from VCR/TV/and so on?" in Chapter 1, "General DVD") appeared worldwide in 2000. This book uses the terms "drive" and "video recorder" to distinguish between recordable computer drives and home set-top recorders.

DVD-RAM is more of a removable storage device for computers than a video recording format, although it has become widely used in DVD video recorders because of the flexibility it provides in editing a recording. The other two recordable format families (DVD-R/RW and DVD+R/RW) are essentially in competition with each other. The market will determine which of them succeeds or if they end up coexisting or merging. There are many claims that one or the other format is better, but they are actually very similar. In 2003 many companies began making drives that could record in both "dash" and "plus" format.

Each writable DVD format is covered briefly below. See Chapter 6's "Hardware and computer components" for hardware manufacturers. For more on writable DVD see Dana Parker's article at www.emediapro.net/EM1999/parker1.html. More information on writable DVD formats is available at industry associations: *RW Products Promotion Initiative* (RWPPI), *Recordable DVD Council* (RDVDC), and DVD+RW Alliance. Also DVD Writers and DVDplusRW.org. If you're interested in writable DVD for data storage, visit Steve Rothman's DVD-DATA page for FAQ and mailing list info.

Is It True There Are Compatibility Problems with Recordable DVD Formats?

Yes. A big problem is that none of the writable formats are fully compatible with each other or even with existing drives and players. In other words, a DVD+R/RW drive can't write a DVD-R or DVD-RW disc, and vice versa (unless it's a combo drive that writes both formats). As time goes by, the different formats are becoming more compatible and more intermixed. A player with the DVD Forum's *DVD Multi* logo is guaranteed to read DVD-R, DVD-RW, and DVD-RAM discs, and a *DVD Multi* recorder can record using all three formats. Some new "Super Multi" drives can write to DVD-R, DVD-RW, DVD+R, and DVD+RW, and even DVD-RAM.

In addition, not all players and drives can read recorded discs. The basic problem is that recordable discs have different reflectivity than pressed discs (the pre-recorded kind you buy in a store—see Chapter 5, "DVD Production"), and not all players have been correctly designed to read them. There are compatibility lists at CustomFlix, DVDMadeEasy, DVDRHelp, YesVideo.com, HomeMovie.com, and Apple web sites that indicate player compatibility with DVD-R and DVD-RW discs. DVDplusRW.org maintains a list of DVD+RW compatible players and drives.

NOTE: Test results vary depending on media quality, handling, writing conditions, player tolerances, and so on. The indications of compatibility in these lists are often anecdotal in nature and are only general guidelines.

Very roughly, DVD-R and DVD+R discs work in about 85 percent of existing drives and players, whereas DVD-RW and DVD+RW discs work in

TABLE 4-2   DVD-R compatibility

	DVD unit	DVD-R(G) unit	DVD-R(A) unit	DVD-RW unit	DVD-RAM unit	DVD+RW unit
DVD-ROM disc	reads	reads	reads	reads	reads	reads
DVD-R(G) disc	often reads	reads, writes	reads	reads, writes	reads	reads
DVD-R(A) disc	usually reads	reads	reads, writes	reads	reads	reads
DVD-RW disc	often reads	reads	reads	reads, writes	usually reads	usually reads
DVD-RAM disc	rarely reads	doesn't read	doesn't read	doesn't read	reads, writes	doesn't read
DVD+RW disc	often reads	usually reads	usually reads	usually reads	usually reads	reads, writes
DVD+R disc	often reads	usually reads	usually reads	routinely reads	reads	reads, usually writes

around 70 percent. The situation is steadily improving. In another few years compatibility problems will mostly be behind us, just as with CD-R (did you know that early CD-Rs had all kinds of compatibility problems?).

Table 4-2 is a summary of recordable DVD compatibility. Below each drive is a column indicating how well it can read or write each format (for simplicity, "doesn't write" is implied if something else is not specified).

DVD-R

DVD-R (which is pronounced "dash R" not "minus R") uses organic dye technology, like CD-R, and is compatible with most DVD drives and players. First-generation capacity was 3.95 billion bytes, later extended to 4.7 billion bytes. Matching the 4.7G capacity of DVD-ROM was crucial for desktop DVD production. In early 2000 the format was split into an "authoring" version and a "general" version. The general version, intended for home use, writes with a cheaper 650-nm laser, the same as DVD-RAM. DVD-R(A) is intended for professional development and uses a 635-nm laser. DVD-R(A) discs are not writable in DVD-R(G) recorders, and vice-versa, but both kinds of discs are readable in most DVD players and drives. The main differences, in addition to recording wavelength, are that DVD-R(G) uses decrementing pre-pit addresses, a pre-stamped (version 1.0) or pre-recorded (version 1.1) control area, CPRM (see "What Are the Copy Protection Issues?" in Chapter 1), and allows double-sided discs. A third version for "special authoring," allowing protected movie content to be recorded on DVD-R media, was considered but will probably not happen.

Pioneer released 3.95G DVD-R(A) 1.0 drives in October 1997 (about 6 months late) for $17,000. New 4.7G DVD-R(A) 1.9 drives appeared in limited quantities in May 1999 (about 6 months late) for $5,400. Version 2.0 drives became available in fall 2000. Version 1.9 drives can be upgraded to 2.0 via downloaded software. (This removes the 2,500 hour recording limit.) New 2.0 [4.7G] media (with newer copy protection features), can only be written in 2.0 drives. 1.9 media (and old 1.0 [3.95G] media) can still be written in 2.0 drives. Version 1.0 (3.95G) discs are still available, and can be recorded in Pioneer DVD-R(A) drives. Although 3.95G discs hold less data, they are more compatible with existing players and drives.

Pioneer's DVR-A03 DVD-R(G) drive was released in May 2001 for under $1000. By August it was available for under $700, and by February 2002 it was under $400. The same drive (model DVR-103) was built into certain Apple Macs and Compaq PCs. Many companies now produce DVD-RW drives, all of which also write CD-R/RW. As of mid-2003 DVD-RW drives are selling for under $200. Most DVD-RAM drives also write DVD-R discs, and some also write DVD-RW discs. Many new drives write both DVD-R/RW and DVD+R/RW.

Pioneer released a professional DVD video recorder in 2002. It sells for about $3000 and provides component video (YPbPr) and 1394 (DV) inputs (along with s-video and composite). It has 1-hour (10 Mbps) and 2-hour (5 Mbps) recording modes, and includes a 2-channel Dolby Digital audio encoder.

Prices for blank DVD-R(A) discs are $10 to $25 (down from the original $50), although cheaper discs seem to have more compatibility problems. Prices for blank DVD-R(G) discs are $2 to $6. Blank media are made by CMC Magnetics, Fuji, Hitachi Maxell, Mitsubishi, Mitsui, Pioneer, Ricoh, Ritek, Taiyo Yuden, Sony, TDK, Verbatim, Victor, and others.

The DVD-R 1.0 format is standardized in ECMA-279. Andy Parsons at Pioneer has written a white paper that explains the differences between DVD-R(G) and DVD-R(A) (www.pioneerelectronics.com/Pioneer/Files/DVDR_whitepaper.pdf).

It's possible to submit DVD-R(A) and DVD-R(G) discs for replication, with limitations. First, not all replicators will accept submissions on DVD-R. Second, there can be problems with compatibility and data loss when using DVD-R, so it's best to generate a checksum that the replicator can verify. Third, DVD-R does not directly support CSS, regions, and Macrovision. Support for this is being added to DVD-R(A) with the *cutting master format* (CMF), which stores DDP information in the control area, but it will take a while before many authoring software programs and replicators support CMF.

DVD-RW

DVD-RW (formerly DVD-R/W and also briefly known as DVD-ER) is a phase-change erasable format. Developed by Pioneer based on DVD-R, using similar track pitch, mark length, and rotation control, DVD-RW is playable in many DVD drives and players. (Some drives and players are confused by DVD-RW media's lower reflectivity into thinking it's a dual-layer disc. In other cases the drive or player doesn't recognize the disc format code and doesn't even try to read the disc. Simple firmware upgrades can solve both problems.) DVD-RW uses groove recording with address info on land areas for synchronization at write time (land data is ignored during reading). Capacity is 4.7 billion bytes. DVD-RW discs can be rewritten about 1,000 times.

In December 1999, Pioneer released DVD-RW home video recorders in Japan. The units cost 250,000 yen (about $2,500) and blank discs cost 3,000 yen (about $30). Since the recorder used the new DVD-VR (video recording) format, the discs wouldn't play in existing players (the discs were *physically* compatible, but not *logically* compatible). Recording time varies from 1 hour to 6 hours, depending on quality. A new version of the recorder was later released that also recorded on DVD-R(G) discs and used the DVD-Video format for better compatibility with existing players.

DVD-RW drives write DVD-R, DVD-RW, CD-R, and CD-RW discs. DVD-RW disc prices are around $5 to $10 (down from the original $30). Blank media are made by CMC Magnetics, Hitachi Maxell, Mitsubishi, Mitsui, Pioneer, Ricoh, Ritek, Sony, Taiyo Yuden, TDK, Verbatim, Victor, and others.

There are three kinds of DVD-RW discs, all with 4.7G capacity. Version 1.0 discs, rarely found outside of Japan, have an embossed lead-in (to prevent copying of CSS information), which causes compatibility problems. Version 1.1 discs have a pre-recorded lead-in that improves compatibility. Version 1.1 discs also come in a "B" version that carries a unique ID in the BCA for use with CPRM. B-type discs are required when copying certain kinds of protected video. (See "What Are the Copy Protection Issues?" in Chapter 1 for more on CPRM; see "What Is BCA?" in Chapter 3, "DVD Technical Details," for more on BCA.)

NOTE: The original Apple SuperDrive (even with older 1.22 firmware) can write to DVD-RW discs, but not from the iDVD application. You must use a different software utility, such as Toast, to write to DVD-RW discs.

DVD-RAM

DVD-RAM, with an initial storage capacity of 2.58 billion bytes, later increased to 4.7, uses phase-change dual (PD) technology with some *magneto-optic* (MO) features mixed in. DVD-RAM is the best suited of the writable DVD formats for use in computers, because of its defect management and zoned CLV format for rapid access. However, it's not compatible with most drives and players (because of defect management, reflectivity differences, and minor format differences). A wobbled groove is used to provide clocking data, with marks written in both the groove and the land between grooves. The grooves and pre-embossed sector headers are molded into the disc during manufacturing. Single-sided DVD-RAM discs come with or without cartridges. There are two types of cartridges: type 1 is sealed; type 2 allows the disc to be removed. Discs can only be written while in the cartridge. Double-sided DVD-RAM discs were initially available in sealed cartridges only, but now come in removable versions as well. Cartridge dimensions are 124.6 mm × 135.5 mm × 8.0 mm. DVD-RAM discs can be rewritten more than 100,000 times, and the discs are expected to last at least 30 years.

DVD-RAM 1.0 drives appeared in June 1998 (about 6 months late) for $500 to $800, with blank discs at about $30 for single-sided and $45 for double-sided. The first DVD-ROM drive to read DVD-RAM discs was released by Panasonic in 1999 (SR-8583, 5x DVD-ROM, 32x CD). Hitachi's GD-5000 drive, released in late 1999, also reads DVD-RAM discs. Blank DVD-RAM media are manufactured by CMC Magnetics, Hitachi Maxell, Eastman Kodak, Mitsubishi, Mitsui, Ritek, TDK, and others.

The spec for DVD-RAM version 2.0, with a capacity of 4.7 billion bytes per side, was published in October 1999. The first drives appeared in June 2000 at about the same price as DVD-RAM 1.0 drives. Single-sided discs were priced around $25, and double-sided discs were around $30. Disc prices were under $10 and retail drive prices were under $200 by 2003. DVD-RAM 2.0 also specifies 8-centimeter discs and cartridges for portable uses such as digital camcorders. Future DVD-RAM discs may use a contrast enhancement layer and a thermal buffer layer to achieve higher density.

Samsung and C-Cube made a technology demonstration (not a product announcement) in October 1999 of a DVD-RAM video recorder using the new DVD-VR format (see the preceding DVD-RW section for more about DVD-VR). Panasonic demonstrated a $3,000 DVD-RAM video recorder at CES in January 2000. It appeared in the U.S. in September for $4,000 (model DMR-E10). At the beginning of 2001, Hitachi and Panasonic released DVD camcorders that use small DVD-RAM discs. The instant access and on-the-fly editing and deleting capabilities of the DVD camcorders are impressive. Panasonic's 2nd-generation DVD-RAM video recorder appeared in October 2001 for $1,500 and also wrote to DVD-R discs.

The DVD-RAM 1.0 format is standardized in ECMA-272 and ECMA-273.

How Do I Remove a DVD-RAM Type 2 Disc from the Cartridge?

Type 2 DVD-RAM cartridges allow the disc to be removed so that it can be played in standard players or drives. (However, most players and drives still aren't able to read the disc; see "Is It True There Are Compatibility Problems with Recordable DVD Formats?")

First break (yes, break) the locking pin by pushing on it with a pointed object such as a ballpoint pen. Remove the locking pin. Unlatch the cover by using a pointed object to press the indentation on the back left corner of the cartridge. Data is recorded on the unprinted side of the disc—do not touch it. When you put the bare disc back the cartridge, make sure the printed side of the shutter and the printed side of the disc face the same direction.

Most DVD-RAM drives will not allow you to write to a bare disc. Some will not allow you to write to a cartridge if the disc has been removed.

DVD+RW and DVD+R

DVD+RW is an erasable format based on CD-RW technology. It became available in late 2001. DVD+RW is supported by Philips, Sony, Hewlett-Packard, Dell, Ricoh, Yamaha, and others. It is not supported by the DVD Forum (even though most of the DVD+RW companies are members), but the Forum has no power to set standards. DVD+RW drives read DVD-ROMs and CDs, and usually read DVD-Rs and DVD-RWs, but do not read or write DVD-RAM discs. DVD+RW drives also write CD-Rs and CD-RWs. DVD+RW discs, which hold 4.7 billion bytes per side, are readable in many existing DVD-Video players and DVD-ROM drives. (They run into the same reflectivity and disc format recognition problems as DVD-RW.)

DVD+RW backers claimed in 1997 that the format would be used only for computer data, not home video, but this was apparently a smokescreen intended to placate the DVD Forum and competitors. The original 1.0 format, which held 3 billion bytes (2.8 gigabytes) per side and was not compatible with any existing players and drives, was abandoned in late 1999.

The DVD+RW format uses phase-change media with a high-frequency wobbled groove that allows it to eliminate linking sectors. This, plus the option of no defect management, allows DVD+RW discs to be written in a way that is compatible with many existing DVD readers. The DVD+RW specification allows for either CLV format for sequential video access (read at CAV speeds by the drive) or CAV format for random access, but CAV recording is not supported by any current hardware. DVD+R discs can only be recorded in CLV mode. Only CLV-formatted discs can be read in standard DVD drives and players. DVD+RW media can be rewritten about 1,000 times (down from 100,000 times in the original 1.0 version).

DVD+R is a write-once variation of DVD+RW, which appeared in mid 2002. It's a dye-based medium, like DVD-R, so it has similar compatibility as DVD-R. Original DVD+RW drives did not fulfill the promise of a simple upgrade to add DVD+R writing support, so they have to be replaced with newer models. The original Philips DVD+RW video recorders, on the other hand, can be customer-upgraded to write +R discs.

Philips announced a DVD+RW home video recorder for late 2001. The Philips recorder uses the DVD-Video format, so discs play in many existing players. HP announced a $600 DVD+RW drive (made by Ricoh) and $16 DVD+RW discs for September 2001. HP's drive reads DVDs at 8x and CDs at 32x, and writes to DVD+RW at 2.4x, CD-R at 12x, and CD-RW at 10x.

In 2003 DVD+R discs cost around $2 to $6 and DVD+RW discs cost around $5 to $10. DVD+RW media are produced by CMC Magnetics, Hewlett-Packard, MCC/Verbatim, Memorex, Mitsubishi, Optodisc, Philips, Ricoh, Ritek, and Sony.

More DVD+RW information is at www.dvdrw.com and www. dvdplusrw.org. The obsolete DVD+RW 1.0 format is standardized in ECMA-274.

Other Recordable Optical Formats

Competitors to recordable DVD were announced but never appeared, thanks in part to the success of the entire DVD family. These formats included AS-MO (formerly MO7), which was to hold 5 to 6 billion bytes, and NEC's *Multimedia Video Disc* (MVDisc, formerly MMVF, *Multimedia Video File*), which was to hold 5.2 billion bytes and was targeted at home recording. ASMO drives were expected to read DVD-ROM and compatible writable formats, but not DVD-RAM. MVDisc was similar to DVD-RW and DVD+RW, using two bonded 0.6mm phase-change substrates, land and groove recording, and a 640nm laser, but contrary to initial reports, the drives were not expected to be able to read DVD-ROM or compatible discs.

There was also FMD (see "What Effect Will FMD Have on DVD?" in Chapter 2). And there are HD formats (see "What's New with DVD Technology?" in Chapter 6, "Miscellaneous").

How Long Does DVD Recording Take?

The time it takes to burn a DVD depends on the speed of the recorder and the amount of data. Playing time of the video may have little to do with recording time, since a half hour at high data rates can take more space than an hour at low data rates. A 2x recorder, running at 22 Mbps, can write a full 4.7G DVD in about 30 minutes. A 4x recorder can do it in about 15 minutes.

NOTE: The -R/RW format often writes a full lead-out to the diameter required by the DVD spec, so small amounts of data (such as a very short video clip) may take the same amount of time as large amounts.

Why Can't I Take a Screenshot of DVD Video? Why Do I Get a Pink or Black Square?

Most DVD PCs, even those with software decoders, use video overlay hardware to insert the video directly into the VGA signal. This an efficient way to handle the very high bandwidth of full-motion video. Some decoder cards, such as the Creative Labs Encore Dxr series and the Sigma Designs Hollywood series, use a pass-through cable that overlays the video into the analog VGA signal after it comes out of the video display card. Video overlay uses a technique called *colorkey* to selectively replace a specified pixel color (often magenta or near-black) with video content. Anywhere a colorkey pixel appears in the computer graphics video, it's replaced by video from the DVD decoder. This process occurs downstream from the computer's video memory, so if you try to take a screenshot (which grabs pixels from video RAM), all you get is a solid square of the colorkey color.

Hardware acceleration must be turned off before screen capture will work. This makes some decoders write to standard video memory. Utilities such as Creative Softworx, HyperSnap, and SD Capture can then grab still pictures. Some player applications such as PowerDVD and the Windows Me player can take screenshots if hardware acceleration is turned off.

Why Can't I Play Movies Copied to My Hard Drive?

Almost all movies are encrypted with CSS copy protection (see "What are the Copy Protection Issues?" in Chapter 1). Decryption keys are stored in the normally inaccessible lead-in area of the disc. You'll usually get an error if you try to copy the contents of an encrypted DVD to a hard drive. However, if you have used a software player to play the movie it will have authenticated the disc in the drive, allowing you to copy without error, but the encryption keys will not be copied. If you try to play the copied VOB files, the decoder will request the keys from the DVD-ROM drive and will fail. You may get the message "Cannot play copy-protected files."

Why Do I Have Problems Playing DVDs on My Computer?

There are thousands of answers to this question, but here are some basic troubleshooting steps to help you track down problems such as jerky playback, pauses, error messages, and so on.

- Get updated software. Driver bugs are the biggest cause of playback problems, ranging from freezes to bogus error messages about regions. Go to the support section on the Web sites of your equipment manufacturers and make sure you have the latest drivers for your graphics adapter, audio card, and DVD decoder (if you have a hardware decoder). Also make sure you have the latest update of the player program.

  Apple has released numerous updates for audio drivers and the DVD player application. Make sure you have the latest versions. Go to the downloads page and search for DVD.

- If you have problems loading a DVD on a Mac, hold down the Command, Option, and I keys when inserting the disk. (This mounts the disc using ISO 9660 instead of UDF.)

- Make sure DMA or SDT is turned on. In Windows, go into the System Properties Device Manager, choose CD-ROM, open the CD/DVD driver properties, choose the Settings tab, and make sure the DMA box (for IDE drives) or the Sync Data Transfer box (for SCSI drives) is checked. Download CD Speed to check the performance of your drive (if it's below 1x, you have problems).

CAUTION: You may run into problems turning DMA on, especially with an AMD K6 CPU or VIA chipset. Check for a BIOS upgrade, a drive controller upgrade, a bus mastering driver upgrade, and a

CD/DVD-ROM driver upgrade from your system manufacturer before turning DMA on. If the drive disappears, reboot in safe mode, uncheck DMA, and reboot again. You may have to tell Windows to restore the registry settings from its last registry backup.

- If you get an error about unavailable overlay surface, reduce the display resolution or number of colors (right-click the desktop and choose the Settings tab).
- Try turning off programs that are running in the background. (In Windows, close or exit applets in the system tray—the icons in the lower right corner. In Mac OS, turn off AppleTalk, file sharing, and virtual memory.)
- Allocate more memory to the Apple DVD Player.
- If you are using a SCSI DVD-ROM drive, make sure that the it's the first or last device in the SCSI chain. If it's the last device, make sure it's terminated.
- Reinstall the Windows bus mastering drivers. (Delete them from the device manager and let Windows ask for the original disc.)
- Bad video when connecting to a TV could be from too long a cable or from interference or a ground loop. See "Why Is the Audio or Video Bad?" in Chapter 3.

Can I Stream DVD over a Network or the Internet?

Short answer: Not if the disc is copy protected.

With a fast enough network (100 Mbps or better, with good performance and low traffic) and a high-performance server, it's possible to stream DVD-Video from a server to client stations. If the source on the server is a DVD-ROM drive (or jukebox), then more than one user simultaneously accessing the same disc will cause breaks in the video unless the server has a fast DVD-ROM drive and a very good caching system designed for streaming video.

A big problem is that CSS-encrypted movies (see "What Are the Copy Protection Issues?" in Chapter 1) can't be remotely sourced because of security issues. The CSS license does not allow decrypted video to be sent over an accessible bus or network, so the decoder has to be on the remote PC. If the decoder has a secure channel to perform authentication with the drive on the server, then it's possible to stream encrypted video over a network to be decrypted and decoded remotely. (But so far almost no decoders can do this.)

One solution is the VideoLAN project which runs on GNU/Linux/Unix, BeOS, Mac OS X, and other operating systems. It includes a player with built-in CSS decryption. Although the code is different from DeCSS, it's an unlicensed implementation and is probably illegal in most countries (see "What is DeCSS?").

An alternative approach is to decode the video at the server and send it to individual stations via separate cables (usually RF). The advantage is that performance is very good, but the disadvantage is that DVD interactivity is usually limited, and every viewer connected to a single drive/decoder must watch the same thing at the same time.

Many companies provide support for streaming video (MPEG-1, MPEG-2, MPEG-4, and so on) over LANs, but only from files or realtime encoders, not from DVD-Video discs.

The Internet is a different matter. It takes over a week to download the contents of a single-layer DVD using a 56k modem. It takes about seven hours on a T1 line. Cable modems theoretically cut the time down to a few hours, but if other users in the same neighborhood have cable modems, bandwidth could drop significantly.

Author's prediction, made in 2001: The average DVD viewing household won't have sufficiently fast Internet connections before 2007 at the earliest. Around that time there will be a new high-definition version of DVD with double the data rate, which will once again exceed the capacity of the typical Internet connection.

What Is DeCSS?

CSS (*Content Scrambling System*) is an encryption and authentication scheme intended to prevent DVD movies from being digitally copied. See "What are the Copy Protection Issues?" in Chapter 1 for details. DeCSS refers to the general process of defeating CSS, as well as to DeCSS source code and programs.

Computer software to decrypt CSS was released to the Internet in October 1999 (see Dana Parker's article at www.emediapro.net/news99/news111.html), although other "ripping" methods were available before that (see "DVD Utilities and Region-free Information" in Chapter 6). The difference between circumventing CSS encryption with DeCSS and intercepting decrypted, decompressed video with a DVD ripper is that DeCSS can be considered illegal under the DMCA and the WIPO treaties. The DeCSS information can be used to "guess" at master keys, such that a standard

PC can generate the entire list of 409 keys, rendering the key secrecy process useless.

In any case, there's not much appeal to being able to copy a set of movie files (often without menus and other DVD special features) that would take over a week to download on a 56K modem and would fill up a 6G hard disk or a dozen CD-Rs. An alternative is to recompress the video with a different encoding format such as DivX (see "What Is Divx?" in Chapter 2) so that it will take less space, but this often results in significantly reduced picture quality. In spite of lower data rates of DivX et al, the time and effort it takes to find and download the files is not worth the bother for most movie viewers. The reality is that most people ripping and downloading DVDs are doing it for the challenge, not to avoid buying discs.

The supporters of DeCSS point out that it was only developed to allow DVD movies to be played on the Linux operating system, which had been excluded from CSS licensing because of its open-source nature. This is specifically allowed by DMCA and WIPO laws. However, the DeCSS.exe program posted on the Internet is a Windows application that decrypts movie files. The lack of differentiation between the DeCSS process in Linux and the DeCSS.exe Windows application is hurting the cause of DeCSS backers, since DeCSS.exe can be used in the process of copying and illegally distributing movies from DVD. See OPENDVD.ORG and Tom Vogt's DECSS CENTRAL for more information on DeCSS.

Worthy of note is that DVD piracy was around long before DeCSS. Serious DVD pirates can copy the disc bit for bit, including the normally unreadable lead in (this can be done with a specially modified drive), or copy the video output from a standard DVD player, or get a copy of the video from another source such as laserdisc, VHS, or a camcorder smuggled into a theater. It's certainly true that DVD piracy is a problem, but DeCSS has little to do with it.

Shortly after the appearance of DeCSS, the DVD CCA filed a lawsuit and requested a temporary injunction in an attempt to prevent web sites from posting (or even linking to!) DeCSS information. The request was denied by a California court on December 29, 1999. On January 14, 2000, the seven top U.S. movie studios (Disney, MGM, Paramount, Sony [Columbia/TriStar], Time Warner, Twentieth Century Fox, and Universal), backed by the MPAA, filed lawsuits in Connecticut and New York in a further attempt to stop the distribution of DeCSS on web sites in those states. On January 21, the judge for the New York suit granted a preliminary injunction, and on January 24, the judge for the CCA suit in California reversed his earlier decision and likewise granted a preliminary injunction. In both cases, the judges ruled that the injunction applied only to sites with DeCSS information, not to linking sites. The CCA suit is based on misappropriation of trade secrets (somewhat shaky ground), while the MPAA suits are based on copyright circumvention. On January 24, 16-year old Jon Johansen, the Norwegian programmer who first distributed DeCSS, was questioned by local police who raided his house and confiscated his computer equipment and cell phone. Johansen says the actual cracking work was done by two anonymous programmers, one German and one Dutch, who call themselves Masters of Reverse Engineering (MoRE).

This all seems to be a losing battle, since the DeCSS source code is available on a T-shirt and was made publicly available by the DVD CCA itself in court records—oops! See Fire, Work With Me (www.brunching.com/copyfire.html) for a facetious look at the broad issue.

How Do I Play DVD Video in HTML, PowerPoint, Director, VB, and So On?

A variety of multimedia development/authoring programs can be extended to play video from a DVD, either as titles and chapters from a DVD-Video volume, or as MPEG-2 files. In Windows, this is usually done with ActiveX controls. On the Mac, until DVD-Video support is added to QuickTime, the options are limited. Newer versions of the Apple DVD Player can be controlled with AppleScript.

DVD-Video and MPEG-2 video can be played back in an HTML page in Microsoft Internet Explorer using many different ActiveX controls (see Table 4-3). Some ActiveX controls also work in PowerPoint, Visual Basic, and other ActiveX hosts. Netscape Navigator is out of the game until it supports ActiveX objects. Simple MPEG-2 playback can be done in PowerPoint using the Insert Movie feature (which requires that a DirectShow-compatible MPEG-2 decoder be installed). DVD and MPEG-2 playback can be integrated into Macromedia Director using specialized Xtras.

Of course, if you simply treat DVD-ROM as a bigger, faster CD-ROM, you can create projects using traditional tools (Director, Flash, Toolbook, Hyper-Card, VB, HTML, and so on) and traditional media types (CinePak, Sorenson, Indeo, Windows Media, and so on in QuickTime or AVI format) and they'll work just fine from DVD. You can even raise the data rate for bigger or better quality video. But it usually won't look as good as MPEG-2.

What Are .IFO, .VOB, and .AOB Files? How Can I Play Them?

The DVD-Video and DVD-Audio specifications (see "Who Invented DVD and Who Owns It? Whom to contact for Specifications and Licensing?" in Chapter 6) define how audio and video data are stored in specialized files.

TABLE 4-3 Active X Controls for DVD Playback

Product	Price	HTML (IE only)	PowerPoint	ActiveX host (VB and so on)	Director
Microsoft MSWebDVD or MSVidWebDVD	free	yes	yes	yes	no
Microsoft Windows Media Player 6.1	free	yes	no	no	no
InterActual PC Friendly	not available	Certain versions	no	no	no
InterActual Player 2.0	$2000 and up	yes	yes	yes	yes
SpinWare iControl	PE: $120, Web: $1200 and up	Web version	PE version	no	no
Visible Light Onstage DVD	$500 and up	ActiveX version	ActiveX version	ActiveX version	Director version
Sonic EDK (InterActual engine)	$4000	yes	no	no	no
Sonic DVD Presenter (InterActual engine)	$40	no	yes	no	no
Tabuleiro DirectMediaXtra	$200	no	no	no	MPEG-2/ VOB files but not DVD-Video volumes
LBO Xtra DVD	$500	no	no	no	yes
Matinée Presenter	?	Separate presentation application. Plays MPEG-2 files (not DVD-Video).			

The .IFO files contain menus and other information about the video and audio. The .BUP files are backup copies of the .IFO files. The .VOB files (for DVD-Video) and .AOB files (for DVD-Audio) are MPEG-2 program streams with additional packets containing navigation and search information.

Since a .VOB file is just a specialized MPEG-2 file, most MPEG-2 decoders and players can play them. You may need to change the extension from .VOB to .MPG. However, any special features such as angles or branching will cause strange effects. The best way to play a .VOB file is to use a DVD player application to play the entire volume (or to open the VIDEO_TS.IFO file), since this will make sure all the DVD-Video features are used properly.

Many DVDs are encrypted, which means the .VOB files won't play when copied to your hard drive. See "Why Can't I Play Movies Copied to My Hard Drive?" in Chapter 4.

If you try to copy the .IFO and .VOB files to a recordable DVD it may not play. See "How Can I Copy a DVD?" in Chapter 5.

You may also run into .VRO files created by DVD video recorders using the -VR format. In some cases you can treat the files just like .VOB files, but in other cases they are fragmented and unplayable. You'll need a utility such as Heuris Extractor to copy them to a hard disk in usable format.

How Do I Get the Microsoft Windows DVD Player Application to Run?

Windows 98 and Windows 2000 included a simple player application. It requires that a DirectShow-compatible DVD decoder be installed (see "Can I Play DVD Movies on My Computer?" in Chapter 4). During setup, Windows installs the player application if it finds a compatible hardware decoder. You must install the player by hand if you want to use it with a software decoder or an unrecognized hardware decoder. Using WinZip or another utility that can extract from cab files, extract dvdplay.exe from driver17.cab (on the original Windows disc). This is the only file you need, but you can also extract the help file from driver11.cab, and you can extract dvdrgn.exe from driver17.cab if you intend to change the drive region.)

Windows Me includes a much improved player, although it still requires a third-party DirectShow-compatible decoder. Windws Me DVD Player is always installed, but it usually does not appear in the Start menu. To use the player, choose Run . . . from the Start menu, then enter dvdplay.

Windows XP moved DVD playback into Windows Media Player. It requires a DVD Decoder Pack, which contains a DirectShow-compatible DVD decoder. See Microsoft's DVD Support in Windows XP page for more info and links to Decoder Packs. Microsoft also has a list of supported software decoders for Windows XP.

I Upgraded to Windows XP. Why Did My DVD Software Stop Working?

DVD player software written for Windows 98 and Me does not work in Windows XP. Most Windows 2000 software also requires an upgrade. Check with your DVD software manufacturer or your PC manufacturer for an upgrade, which in many cases is free. Or you may want to buy a low-cost Windows XP DVD Decoder Pack (see the preceding "How Do I Get the Microsoft Windows DVD Player Application to Run?").

How Can I Rip Audio from a DVD to Play as MP3 or Burn to a CD?

Keep in mind that unless you are copying something for your own personal use from a DVD that you own, copying a DVD is usually a copyright violation, which is illegal and dishonest.

Use a DVD ripping tool (see "What is DeCSS?" in Chapter 4 and "DVD Utilities and Region-free Information" in Chapter 6) to extract Dolby Digital or PCM (WAV) files from a DVD. Then use a utility to convert to MP3, WMA, or other formats, or to burn to an audio CD.

Alternatively you can connect the audio output from a DVD player (see "What Are the Outputs of a DVD Player?" in Chapter 3) to an audio recorder or to audio inputs on a computer.

Chapter 5

DVD Production

DVD production has two basic phases: *development* and *publishing*. Development is different for DVD-ROM and DVD-Video, publishing is essentially the same for both. Cheap, low-volume productions can be *duplicated* on recordable discs, whereas high-volume, mass-market products such as movies must be *replicated* in specialized factories.

DVD-ROM content can be developed with traditional software development tools such as Macromedia Director, Visual BASIC, Quark mTropolis, or C++. Discs, including DVD-R check discs, can be created with UDF formatting software (see "What DVD-ROM Formatting Tools Are Available?"). DVD-ROMs that take advantage of DVD-Video's MPEG-2 video and multichannel Dolby Digital or MPEG-2 audio require video and audio encoding (see "What DVD Production Tools are Available?").

DVD-Video content development has three basic parts: *encoding*, *authoring* (design, layout, and testing), and *premastering* (formatting a disc image). The entire development process is sometimes referred to as authoring. Development facilities are provided by many service bureaus (see "Who Can Produce a DVD for Me?"). If you intend to produce numerous DVD-Video titles (or you want to set up a service bureau), you may want to invest in encoding and authoring systems (see "What DVD production tools are available?" and "What DVD Authoring Systems Are Available?").

Replication (including mastering) is the process of "pressing" discs in production lines that spit out a new disc every few seconds. Replication is done by large plants (see "Who Can Produce a DVD for Me?" for a list) that also replicate CDs. DVD replication equipment typically costs millions of dollars. A variety of machines are used to create a glass master, create metal stamping masters, stamp substrates in hydraulic molds, apply reflective layers, bond substrates together, print labels, and insert discs in packages. Most replication plants provide "one-off" or "check disc" services, where one to a hundred discs are made for testing before mass duplication. Unlike DVD-ROM mastering, DVD-Video mastering may include an additional step for CSS encryption, Macrovision, and regionalization. There is more information on mastering and replication at Technicolor and Disctronics.

For projects requiring fewer than 50 copies, it can be cheaper to use recordable discs (see "What About Recordable DVD: DVD-R, DVD-RAM, DVD-RW, DVD+RW, and DVD+R?" in Chapter 4). Automated machines can feed recordable blanks into a recorder, and even print labels on each disc. This is called *duplication*, as distinguished from replication.

How Much Does It Cost to Produce a DVD? Isn't It More Expensive than Videotape, Laserdisc, and CD-ROM?

Videotape, laserdisc, and CD-ROM can't be compared to DVD in a straight-forward manner. There are basically three stages of costs: production, pre-mastering (authoring, encoding, and formatting), and mastering/replication.

DVD video production costs are not much higher than for VHS and similar video formats unless the extra features of DVD such as multiple sound tracks, camera angles, seamless branching, and so on are employed.

Authoring and pre-mastering costs are proportionately the most expensive part of DVD. Video and audio must be encoded, menus and control information have to be authored and encoded, it all has to be multiplexed into a single data stream, and finally encoded in low level format. Typical charges for compression are $50/min for video, $20/min for audio, $6/min for subtitles, plus formatting and testing at about $30/min. A ballpark cost for producing a Hollywood-quality two-hour DVD movie with motion menus, multiple audio tracks, subtitles, trailers, and a few info screens is about $20,000. Alternatively, many facilities charge for time, at rates of around $300/hour. A simple two-hour DVD-Video title with menus and various video clips can cost as low as $2,000. If you want to do it yourself, authoring and encoding systems can be purchased at prices from $50 to over $2 million. See "How Do I Copy my Home Videos/Movies/Slides to DVD?" for more on low-cost DVD creation.

Videotapes don't really have a mastering cost, and they run about $2.40 for replication. CDs cost about $1,000 to master and $0.50 to replicate. Laserdiscs cost about $3,000 to master and about $8 to replicate. As of 2003, DVDs cost about $1000 to master and about $0.70 to replicate. Double-sided or dual-layer discs cost about $0.30 more to replicate, since all that's required is stamping data on the second substrate (and using transparent glue for dual layers). Double-sided, dual-layer discs (DVD-18s) are more difficult and more expensive (see "When Did Double-sided, Dual-layer Discs (DVD-18) Become Available?" in Chapter 3, "DVD Technical Details").

What DVD-ROM Formatting Tools Are Available?

- Ahead (www.nero.com)

  - *Nero*. DVD formatting software for Windows. Can make disc image files and bootable discs. $70.

- GEAR (www.gearcdr.com)

  - *GEAR Pro DVD*. DVD formatting software for Windows 95/98/NT4. Writes to DVD-R, DVD-RAM, jukeboxes, and tape, along with general UDF formatting and CD-R/RW burning features. $700.

  - *DVD RomMaker*. DVD formatting systems with RAID hardware. $60,000 to $100,000.

- Philips (www.philips.com)

  - *DVD-ROM Disc Builder*. DVD formatting software for Windows NT. Writes to tape.

- Roxio (www.roxio.com)

  - *Toast DVD*. DVD formatting software for Mac OS. Writes to DVD-R and tape. Can create DVD-Video discs from VOB and IFO files. $200

- SmartDisk (www.smartdisk.com, acquired MTC)

  - *ForDVD*. DVD formatting software for Windows. Writes to DVD-R and tape. Can create DVD-Video discs from VOB and IFO files. $1200.

- Smart Storage (www.smartstorage.com)

  - *SmartDVD Maker*. DVD formatting software for Windows NT. Writes to DVD-R and tape. Can create DVD-Video discs from VOB and IFO files. $2500. (Discontinued as of March 2001.)

- Software Architects (www.softarch.com)

  - *WriteDVD Pro* and *WriteUDF*. DVD formatting software for Mac OS and Windows. Writes to DVD-R and DVD-RAM.

- Sonic (www.sonic.com, acquired Daikin and Veritas DMD)

  - *DVD-ROM Formatter*. DVD formatting software for Windows NT/2000/XP. Writes to DVD-R and tape. Can create DVD-Video discs from VOB and IFO files.

- Stomp (www.stompinc.com, retail distributor for certain Sonic products)
  - *RecordNow* and *MaxRecordNow MAX Platinum*. CD and DVD burning software for music, photos, and video. Windows. $50 and $80.
  - *Backup MyPC* and *Simple Backup*. Windows file backup software for recordable DVD and CD.
- Veritas (acquired Prassi)

Veritas Desktop and Mobile Division was acquired by Sonic in November 2002. Veritas products such as RecordNow and Drive Letter Access are now from Sonic, distributed by Stomp.

- Young Minds (www.ymi.com)
  - *DVD Studio* and *MakeDisc for DVD*. DVD formatting software for Windows NT and Unix. Writes to DVD-R.

Features to Look for in DVD Formatters:

- Support for UDF file system, including MicroUDF (UDF 1.02 Appendix 6.9) for DVD-Video and DVD-Audio zones.
- Support for UDF bridge format, which stores both UDF and ISO-9660 file systems on the disc.
- Ability to recognize VIDEO_TS and AUDIO_TS directories (containing IFO, VOB, and AOB files) and place them contiguously at the physical beginning of the disc for compatibility with DVD-Video players. Placement of directory entries in the first UDF file descriptor is also needed for compatibility with certain deficient consumer players.
- Support for long filenames in Windows (Joliet format recommended).
- Full equivalence between UDF and Joliet (ISO-9660) filenames. (Windows NT 4.0 and Windows 98 read Joliet filenames; Mac OS 8.1+, Windows 98, and Windows 2000 read UDF filenames. MS-DOS and Windows 95 and earlier read ISO-9660 filenames. Mac OS 8.0 and earlier read HFS or ISO-9660 filenames.)
- Proper truncation and translation of ISO-9660 filenames to 8.3 format for discs intended for use with MS-DOS and certain other OSs.
- Support for Mac OS file information within the UDF file system (for use with Mac OS 8.1 and later).
- Support for Mac OS HFS file system if icons and other file information is needed for Mac OS versions earlier than 8.1.

- Ability to create a bootable disc using the El Torito specification in the ISO-9660 sectors.

What DVD Production Tools Are Available?

Video Encoding Tools

- Brent Beyeler (various download sites)
  - *bbMPEG*. Basic MPEG-2 encoder for Windows. Free.
- Canopus (www.canopus.com)
  - *ProCoder*. Software video format converter with MPEG encoding. Two-pass VBR. Advanced features such as NTSC-PAL conversion, de-interlacing, 2-3 pulldown, and batch processing. Windows. $700.
  - *MVR1000*. Hardware real-time video capture and MPEG encoder board for Windows. VBR and CBR. Includes Sonic *DVDit SE* for DVD/VCD authoring.
  - *Amber*. MPEG-2 hardware designed for encoding and archiving video in MPEG format. VBR and CBR. (Panasonic MN85560 encoder chip). Windows. $2,000.
  - *DVRaptor RT*. Hardware DV video editing with MPEG output. Windows. $600.
  - *DVStorm*. Hardware video editing/encoding system for MPEG and DV. Includes Ulead *DVD Workshop* for DVD/VCD authoring. Windows. $1,100.
  - *DVRex RT Professional*. Hardware video editing/encoding system for MPEG and DV. Includes Sonic *DVDit SE* for DVD/VCD authoring. Windows. $4,400.
- Custom Technology (www.cinemacraft.com)
  - *Cinemacraft Encoder-PRO*. MPEG-2 real-time NTSC video encoding software for Windows NT. $38,000.
  - *Cinema Craft Encoder-SP*. MPEG video encoding software for Windows XP and 2000. CBR and VBR. $2,000.
  - *Cinema Craft Encoder-Basic*. MPEG video encoding software for Windows. CBR and VBR. $60.
- Darim (www.darim.com)
  - MPEGator 2. MPEG-2 real-time encoding hardware for Windows and Windows NT. $1,800.

- Dazzle (www.dazzle.com)
  - *Digital Video Creator II*. MPEG-2 video capture/edit/encode system with PCI card. Includes Sonic *DVDit LE*. Windows 98/2000. $300.
- Digital Ventures (www.dvdcomposer.com)
  - *DVDComposer*. MPEG-2 video encoding system for SGI. VBR and CBR. (C-Cube chip). $50,000.
- Digital Vision (www.digitalvision.se)
  - *BitPack*. MPEG-2 video encoding workstation. Extendable to HDTV.
  - DVNR system for video preprocessing.
- Digigami (www.digigami.com)
  - *MegaPeg*. MPEG-2 video encoding software for Windows. VBR and CBR. $500. Also available as Adobe Premiere plug-in for Windows or PowerMac. $400.
- DV Studio (www.dv-studio.com)
  - *Apollo Expert*. MPEG-2 video encoding (and decoding) hardware for Windows NT. $2,000.
- FlaskMPEG (www.flaskmpeg.com)
  - Freeware encoding software for Windows.
- Gunjarm Digital (www.gunjarm.co.kr)
  - *MPEGRich*. Professional MPEG-2 real-time encoding hardware. CBR and VBR. Windows NT.
- Heuris (www.heuris.com)
  - MPEG Power Professional 1, MPEG Power Professional 2, MPEG Power Professional DVD, MPEG Power Professional DTV-SD, and Power Professional DTV-HD. MPEG-2 video encoding software for Mac OS and Windows. DVD and DTV versions include VBR encoding. $350, $1,000, $1,500 and $2,500.
  - Cyclone. MPEG-1 and MPEG-2 encoding software designed for OEMs. Mac OS and Windows NT.
- Ligos (www.ligos.com)
  - LSX-MPEG Encoder. MPEG-2 video encoding software. CBR and VBR. Windows. $150.
  - LSX-MPEG Suite. Adobe Premiere plug-in for producing MPEG-1 or MPEG-2 output. Includes standalone LSX-MPEG player. Windows 9x/NT. $400.

- Media100 (www.media100.com)
  - iFinish RealTime MPEG Option. Editing software with MPEG-2 video encoding add-on. Windows NT. $6,000 to $18,000.
- Microcosmos/Nanocosmos (www.nanocosmos.de)
  - MPEG SoftEngine. MPEG-2 video encoding software for Windows, Solaris, and Linux. $250 to $3500.
- Minerva (www.minervasys.com)
  - Compressionist 110, 200, and 250. Professional MPEG-2 real-time encoding hardware. CBR and VBR. Mac OS host computer. $70,000. [No longer available.]
  - Publisher 300. Professional MPEG-2 video and MPEG Layer 2 audio real-time encoding hardware. CBR and VBR. Mac OS. [No longer available.]
- Optibase (www.optibase.com)
  - MPEG MovieMaker 200. Professional MPEG-2 video and Dolby Digital audio real-time encoding hardware for Windows and Windows NT. CBR and VBR. $7,000 to $22,000.
- Philips (www.philips.com)
  - *DVS3110*. Professional MPEG-2 video encoder for PAL and NTSC. CBR and VBR.
- PixelTools (www.pixeltools.com)
  - *Expert-DVD*. MPEG-2 video encoding software. CBR and VBR. Windows. $2,000.
  - *Simple-DVD*. AVI-to-DVD conversion utility for Windows. $1,500.
- Snell & Wilcox (www.snellwilcox.com)
  - *Prefix CPP100, Prefix CPP200, NRS500, Kudos NRS50,* and *Kudos NRS30.* Video preprocessors (noise reduction and image enhancement).
- Sonic Solutions (www.sonic.com)
  - *SD-1000*. Professional MPEG-2 video encoding hardware. CBR and VBR. Segment-based reencoding. Mac OS and Windows OS. $13,000.
  - *DVD Fusion*. Encoding/authoring plug-in for Media 100 and QuickTime video editing systems. Hardware-accelerated version (velocity engine) encodes VBR and CBR in real time. Mac OS. $8,000 and $12,000.

- Sony (www.sony.com/professional)
  - *DVA-V1100*. High-end MPEG-2 video encoding hardware. CBR and VBR. Windows NT.
- Spruce Technologies (www.spruce-tech.com)

Spruce was acquired in July 2001 by Apple. The MPX-3000 encoder will continue to be sold by dealers.

  - *MPX-3000*. Professional MPEG-2 real-time encoding hardware. CBR and VBR. Windows NT.
  - *MPEGXpress 2000* (formerly from CagEnt). Professional MPEG-2 real-time encoding hardware. CBR and VBR. Windows NT.
- TMPGEnc (www.tmpgenc.net)
  - *TMPGEnc* and *TMPGEnc Pro*. MPEG-1 and MPEG-2 software video encoders, plus multiplexing/demultiplexing, file joining, and trimming tools. Free.
- VisionTech (www.visiontech-dml.com)
  - *MVCast*. Low-end real-time MPEG-2 video/audio encoding hardware for Windows NT and Solaris. AVI-to-MPEG-2 conversion. $2000.
- Vitech (www.vitecmm.com)
  - *MPEG Toolbox-2*. AVI to MPEG-2 VBR/CBR. MPEG-2 video editing. Windows 95/98/NT. $250.
- Wired (www.wiredinc.com)
  - *MediaPress*. MPEG-2 encoding hardware (PCI). CBR and VBR. Mac OS and Windows 95/98/NT. $2,500.
- Zapex (www.zapex.com)
  - *ZP-200*. Real-time PCI encoder for MPEG-2 video and PCM Audio. Non-real-time encoding and VOB multiplexing from Adobe Premiere. Windows NT.
  - *ZP-300*. Real-time PCI Encoder for CBR/VBR MPEG-2 video, 2-channel Dolby Digital, and PCM Audio. Non-real-time encoding and VOB multiplexing from Adobe Premiere. Windows NT.

Audio Encoding Tools

- Digital Vision (www.digitalvision.se)
  - *BitPack*. Multichannel audio encoding workstation for Dolby Digital, MPEG-2, and PCM.

- Dolby (www.dolby.com)
  - *DP569*. Multichannel Dolby Digital audio encoding hardware.
- Kind of Loud Technologies (www.uaudio.com)
  - *SmartCode Pro/Dolby Digital*. 5.1-channel encoding software plugin for Digidesign Pro Tools. $1000.
  - *SmartCode Pro/DTS*. 5.1-channel encoding software plugin for Digidesign Pro Tools. $2000.
- Microcosmos (now www.nanocosmos.com)
  - *MPEG SoftEngine/Audio*. MPEG audio encoding software for Windows/Solaris. $95/$350.
- Minerva (www.minervasys.com)
  - *Audio Compressionist*. Professional Dolby Digital real-time, 5.1-channel encoder. Windows NT.
- Minnetonka Audio Software (www.minnetonkaaudio.com)
  - *SurCode for Dolby Digital*. Multichannel Dolby Digital audio encoding software. $1000.
  - *SurCode DVD Professional for DTS*. Multichannel DTS audio encoding software. $2000.
- Philips (www.philips.com)
  - *DVD3310*. Professional MPEG-2 multichannel audio encoder.
- PixelTools (www.pixeltools.com)
  - *Expert-Audio*. MPEG Layer 2 audio encoding software. Windows.
- Sonic Solutions (www.sonic.com)
  - *Sonic DVD Studio*. Professional real-time Dolby Digital 5.1, MPEG-2, and PCM audio encoding hardware. Mac OS.
  - *MLP Encoder*. $9,000.
- Sonic Foundry (www.sonicfoundry.com)
  - *Soft Encode*. Dolby Digital 2-channel or 5.1-channel audio encoding software. Windows 95/98/NT. $500 (2 channels) or $900 (5.1 channels).
- Sony (www.sony.com/professional)
  - *DVA-A1100*. High-end, real-time Dolby Digital 5.1, MPEG-2, and PCM audio encoding hardware. Windows NT.
- Spruce Technologies (www.spruce-tech.com)

Spruce was acquired in July 2001 by Apple. The ACX-5100 encoder will continue to be sold by dealers.

- *ACX- 5100* (formerly from CagEnt). Professional Dolby Digital real-time, 5.1-channel encoder. Windows NT.
- *ACX-2000* (formerly from CagEnt). Professional Dolby Digital real-time, 2-channel encoder. Windows NT.

- Zapex (www.zapex.com)
  - *ZP-100*. Real-time PCI encoder for 2- or 5.1-channel Dolby Digital and MPEG Layer 2. Windows NT.

Other Production Tools

- Alcohol Software (www.alcohol-software.com)
  - *Alcohol 52%*. Emulate CDs and DVDs without physical disc. Windows. $28.
  - *Alcohol 68%*. Copy CDs and DVDs. Windows. $30.
  - *Alcohol 120%*. Combination of *Alcohol 52%* and *Alcohol 68%*. Windows. $50.

- ASINT (www.asintgroup.com)
  - Industrial DVD players, touchscreens, and DVD kiosk products.

- BCD Associates (www.bcdusa.com)
  - DVD controllers for custom installations.

- Cambridge Multimedia (www.cmgroup.co.uk)
  - Touchscreens and other custom interfaces for industrial DVD players.

- Computer Prompting & Captioning Co. (www.cpcweb.com)
  - *CPC-DVD*. Closed Caption production system. DOS. $6,000.

- DCA (Doug Carson & Associates, www.dcainc.com)
  - *MIS (Mastering Interface System)*. Mastering interface system for DVD and CD. Windows NT.
  - *ITS (Image Transfer System)*. Transfer and convert DVD and CD images.
  - *DVS+* (Data Verification System). Checks DVD and CD images. Includes Interra *Surveyor* to check for DVD-Video spec compliance. Can transfer between discs and tape. Windows NT.

- *INMS (Integrated Network Mastering System)*. Combination of MIS, ITS, DVS+ in a system with a RAID.

- Eclipse Data Technologies (www.eclipsedata.com)
  - *EclipseSuite*. DVD and CD premastering tools to copy and verify images, copy tapes, and so on. Windows NT.
  - *ImageEncoder*. LBR mastering interface for CD and DVD mastering. Windows NT.

- FAB (www.fab-online.com)
  - *FAB Subtitler DVD Edition*. Subtitle generator program (text and bitmap formats) that works with most DVD authoring systems. Windows.

- Heuris (www.heuris.com)
  - *Xtractor*. Software to extract video and audio streams from unencrypted DVD-Video discs and DVD-VR discs. $150.

- Isomedia (www.isomedia.com)
  - DVD DLT utilities: copy DLTs, extract images, inspect ISO/UDF/DDP info, checksums, and so on.

- Museum Technology Source (www.museumtech.com)
  - DVD controllers for Pioneer industrial players in custom installations.

- Novastor (www.novastor.com)
  - *TapeCopy*. Copy DLTs, inspect tape blocks.

- PixelTools (www.pixeltools.com)
  - *MPEGRepair*. Software to analyze, repair, insert Closed Captions, add panscan vectors, and do other handy things to MPEG files. Windows.

- Smart Projects (www.ping.be/vcd/isobuster.htm)
  - *ISOBuster*. Inspect CD and DVD volumes and image files. Free.

- SoftNI (www.softni.com)
  - *The DVD Subtitler*. Subtitle graphics preparation software. Windows 95/98/NT/2000.
  - *The Caption Encoder*. Closed Caption production system. DOS, Windows 95/98.
  - *The Caption Retriever*. Closed Caption recovery and decoding system. Windows 95/98/NT/2000.

- Tapedisk (www.tapedisk.com)
  - *TD Raw*. Reads raw data from a SCSI tape drive as if it were a hard disk. DOS/Windows. $500.
  - *TD RAW NT*. Version of TD Raw for Windows NT 4.0. $750.
- Technovision (www.technovision.com)
  - Touchscreens and other custom interfaces for industrial DVD players.
- Teco (www.tecoltd.com)
  - *ParseMPEG* ($500) and *Bitrate Viewer* (free). Software to analyze MPEG streams. Windows.

Also see "What Testing/Verification Services and Tools Are Availble?" for DVD emulation, verification, and analysis tools.

Other Production Services

- Captions, Inc. (www.captionsinc.com, Burbank, CA), 818-729-9501. Captioning and subtitle services.
- European Captioning Institute (ECI) (www.ecisubtitling.com, London, UK). +44 (0)20 7323 4657. Captioning and subtitle services.
- Captioneering (www.captioneering.com, Burbank, CA), 888-418-4782. Captioning and subtitle services.
- National Captioning Institute (NCI) (www.ncicap.org, LA 818-238-4201; NY 212-557-7011; VA 703-917-7619). Captioning and subtitle services.
- SDI Media Group (www.sdi-media.com, worldwide), +44 (0)20 7349. Subtitle services.
- Softitler (www.softitler.com, Los Angeles, CA). Subtitle services.
- Tele-Cine (www.telecine.co.uk, London, UK), +44 (0) 171 208 2200. Film-to-video conversion.
- TelecineMojo (www.telecinemojo.com, Los Angeles, CA), 323-697-0695. Film-to-video conversion.
- Vitac (www.vitac.com, Canonsburg, PA) 888-528-4822. Captioning services.

What DVD Authoring Systems Are Available?

For more detail on the systems listed below, follow the links or see the comparison table of selected DVD authoring systems at DVDirect.

- Apple (www.apple.com)
  - *DVD Studio Pro*. Mid-level DVD-Video authoring tool for Mac OS. $1,000.
  - *iDVD*. Simple, drag-and-drop DVD-Video authoring, bundled with Macs that have DVD-R drives.
  - *DVDMaestro*. Windows. See Spruce, below.

- Astarte (www.astarte.de)

Astarte was acquired April 2000 by Apple, so their products are generally no longer available. They resurfaced in March 2001 as iDVD and DVD Studio Pro from Apple.

  - *DVDirector* and *DVDirector Pro*. Low-end and mid-level DVD-Video authoring tools for Mac OS. Pro version includes *MediaPress* hardware MPEG-2 encoder from Wired. *Millennium Bundle* turnkey workstation includes *DVDirector Pro*, Mac G4, and more. $5,400, $10,00, $15,000.
  - *DVDelight*. Simple, drag-and-drop DVD-Video authoring for Mac OS. $1,000.
  - *DVDExport*. Software to convert Macromedia Director presentations to DVD-Video format. Mac OS. $900.

- Authoringware (www.authoringware.com)
  - *DVD WISE*. Mid-level authoring system for Windows 95/98/NT. $950.
  - *DVD Quickbuilder*. Multiplexing software.

- Avid (www.avid.com)
  - *Xpress DV*. Video editing software with DVD-Video output (using Sonic AuthorScript). $1,700.
  - *Xpress DV Powerpack*. *Xpress DV* with other software, including Sonic DVDit SE. $3,000.

- Blossom Technologies (www.blossomvideo.com)
  - *DaViD 2000, 4000, 6000, and 10000*. Turnkey Windows NT 4.0 systems using Daikin *Scenarist* authoring software and Optibase encoding hardware or Sonic Foundry audio encoding software. $20,000 to $100,000.

- Canopus (www.canopus.com)
  - *Amber for DVD*. Amber MPEG-2 encoding hardware with Spruce *DVDVirtuoso* authoring software. $3,300.

- Daikin (www.sonic.com/products/scenarist, Daikin US Comtec Laboratories)

Daikin's DVD business was acquired by Sonic in February 2001. Scenarist, ReelDVD, and ROM Formatter are now carried by Sonic.

- DV Studio (www.dv-studio.com)

  - *Apollo Expert Author* and *Apollo Expert DVDer*. Mid-level DVD-Video authoring system for Windows NT, using DV Studio *Apollo Expert* MPEG-2 encoding hardware and Intec *DVDAuthorQuick* authoring software (*Author* package, $4,000) or Sonic *DVDit* (*DVDer* package, $2,500).

  - *Apollo Expert Archiver*. MPEG-2 encoding system for archiving video to DVD-RAM. $2,500 (DVD-RAM drive included).

- Futuretel (www.futuretel.com)

  - *Crescendo.*

- Gunjarm Digital (www.gunjarm.co.kr)

  - *DVDRich*. Mid-level DVD-Video authoring/encoding on Windows NT. Uses MPEGRich encoder and Daikin Scenarist or Intec DVDAuthorQuick. $30,000.

- Houpert Digital Audio (HDA, www.cube-tec.com)

  - *CubeDVD-A*. DVD-Audio authoring plug-in module for Cube-Tec AudioCube digital audio workstation. Uses audio assets mastered by NuendoCube. Windows 2000.

- Intec America

  - *DVDAuthorQuick*. Mid-level and low-level DVD-Video authoring software line for Windows NT. Comes in three versions: *Pro*, *Desktop*, and *LE*. $8,000, $2,500, and $400.

- Margi Systems (www.margi.com)

  - *DVPublish-to-Go*. Simple DVD authoring to DVD-R or CD-R/W. Includes Margi's *1394-to-Go* PC, MGI's *VideoWave III*, and Sonic Solution's *DVDit LETx*. Windows 98 SE or 2000. $300.

- Matrox (www.matrox.com)

  - *Matrox RT2000* and *DigiSuite DTV*. Video capture and editing in DV and MPEG-2 formats. Includes Sonic Solution's *DVDit LE* for simple DVD authoring. Windows 98. $1,300.

- Microboards (www.microboards.com)

  - *DVD AuthorSuite.* DVD-Video authoring/encoding for Windows NT. Uses Intec *DVDAuthorQuick* software, Zapex encoders, and Sigma Designs decoder. $25,000.

- Minerva (www.minervasys.com)

Minerva DVD authoring software was acquired by Pinnacle in 2000, so it is no longer generally available. Impression was re-released by Pinnacle in July 2001.

  - *DVD-Professional SL* and *DVD-Professional XL.* DVD-Video authoring/encoding systems for Windows NT. Includes Publisher 300 and Minerva Studio. $100,000.

  - *Impression.* DVD-Video authoring/encoding system for Windows. $10,000.

- Minnetonka Audio Software (www.minnetonkaaudio.com)

  - *DiscWelder Steel.* Basic DVD-Audio authoring software. Windows. $500.

  - *A-Plus.* Basic DVD-Audio authoring software. Windows. $2,000.

  - *DiscWelder Chrome.* Professional DVD-Audio authoring software. Windows. $3500.

- MTC (www.mtc2000.com, Multimedia Technology Center)

MTC was acquired by SmartDisk in 2000.

  - *StreamWeaver Express* and *StreamWeaver Pro.* Simple and mid-level DVD-Video authoring, and $900 premastering on Windows. $900 and $3,000.

  - *DVDMotion.* Authoring systems for Windows, oriented toward multimedia DVD-ROM production. Comes in three versions: *Pro*, *SE* (Standard), *CE* (Consumer). $1,000, $400, $95.

  - *DVDMotion CE.* Entry-level authoring system for Windows 98/NT4. $75.

- NEC (www.nec-global.com)

  - *DV Editor.* IEEE-1394 card and and software, plus Sonic's *DVDit LE.* Windows 98. Available in Japan only.

  - Optibase (www.optibase.com)

- *DVD-Fab XPress* and *DVD-Fab*. Turnkey DVD-Video authoring/encoding systems for Windows NT. Includes Optibase *MPEG Fusion* MPEG-2 encoder and Daikin *Scenarist* authoring software. $35,000.

- Panasonic (http://matsushita.co.jp/info/mei.html)

  - *LQ-VD2000S*. Turnkey professional DVD-Video authoring system, including Windows NT 4.0 workstation. Uses Panasonic MPEG-2 encoder and *LQ-VD3000* emulator. $120,000.

  - *LQ-VDS120*. Additional workstation software for networking with *LQ-VD2000S*. $22,550

  - *LQ-VD3000*. DVD Emulator. $15,000

- Pinnacle (www.pinnaclesys.com)

  - *DVD1000*. MPEG-2 video editing and DVD-Video authoring system for Windows. Pinnacle DVD1000 hardware with Adobe Premiere and Minerva Impression. $8,000.

  - *Impression DVD*. Mid-level DVD-Video authoring/encoding system for Windows. $1,000.

  - *Pinnacle Pro-ONE*. DVD editing/authoring package. Uses Adobe *Premiere* and *Impression DVD-SE*. $1,300.

  - *Pinnacle Edition*. Video editing with linear-play DVD/SVCD output. $700.

- Philips (www.philips.com)

  - *DVD-Video Disc Designer* and *DVD-Video Authoring Toolset*. Windows NT.

- PixelTools (www.pixeltools.com)

  - DVDPlug-in. Simple encoding/authoring plug-in for Adobe Premiere. Generates single-program, autoplay images that can be copied to recordable discs.

- Q-Comm

  - *EasyDVD*

- Roxio (www.roxio.com)

  - *Easy CD and DVD Creator*. Basic VCD and DVD creation for Windows. $80.

- SADiE (www.sadie.com)

  - *DVD-A Direct*. DVD-Audio authoring package for SADiE editing and mastering system. Windows.

- Sonic Solutions (www.sonic.com)

  - *Scenarist SGI*. DVD-Video authoring for SGI. The original professional system. $25,000.

  - *Scenarist NT*. Professional DVD-Video authoring on Windows NT. Comes in two versions: *Advanced*, $15,000; *Professional*, $22,000.

  - *DVD Creator*. Professional DVD-Video authoring/encoding systems for corporate and industrial applications. Mac OS. Various configurations: *DVD Creator All-in-One Workstation*, $80,000; *DVD Creator Encoding*, $24,500, *DVD Creator Authoring*, $15,000.

  - *DVD-Audio Creator*. DVD-Audio authoring system (co-developed with Panasonic). Windows. *DVD-Audio Complete Workgroup*, $53,000; *DVD-Audio Creator*, $13,000; *DVD-Audio Creator LE*, $6,000.

  - *OneClick DVD*. Simple DVD-Audio authoring. Mac OS. $15,000.

  - *DVD Fusion*. Mid-level DVD-Video authoring system. Mac OS.

  - *DVD Producer* (formerly *DVD Fusion for Windows*). Mid-level DVD-Video authoring system. Windows. $3,000.

  - *ReelDVD*. Low-end authoring for NT and Windows 2000. $1,500.

  - *DVDit LE* (limited), *SE* (standard), and *PE* (professional). Simple, drag-and-drop DVD-Video authoring for Windows. $300 (SE), $400 (PE). *LE* version is designed to be bundled with other hardware and software products.

  - *MyDVD*. Simple personal DVD-Video authoring for Windows. $79. Generally bundled with DVD recorders.

- Sony (www.sony.com/professional)

  - *DVA-1100*. High-end authoring/encoding system with one to eight stations. Price range starts at $175,000.

- Spruce Technologies (www.spruce-tech.com)

Spruce was acquired in July 2001 by Apple. DVDMaestro may still be available from some dealers.

  - *DVDMaestro*. High-end authoring/encoding systems for Windows NT. $25,000.

  - *DVDConductor, DVDVirtuoso, DVDPerformer*. Mid-level authoring/encoding systems for Windows NT. Also allow DVD content to be recorded and played from CD-R. $10,000, $1500, $?.

  - *SpruceUp*. Simple personal DVD-Video authoring for Windows (NT4/98/ME/NT/2000). $129.

  - *DVDStationCX*. Turnkey system using DVDConductor. $25,000.

- *DVDStationNLE*. Turnkey system using DVDConductor and Heuris *MPEG Power Professional* encoding software. $10,000.
- *DVDTransfer*. Turnkey automated tape-to-DVD system. $30,000.

- Ulead (www.ulead.com)
  - *MediaStudio Director's Cut*. Video editing software with built-in DVD authoring. $190.
  - *DVD Workshop*. Basic video editing and DVD authoring. $300.

- Visible Light (www.visiblelight.com)
  - *Macarena* and *Macarena Pro*. DVD-Video authoring for Power Mac G4. Software encoding or hardware encoding (Pro version). Uses Astarte *DVDirector* software. $10,000 and $15,000.

- Vitech (www.vitecmm.com)
  - *DVD Toolbox*. AVI to DVD-Video. Write to CD-R, DVD-R, DVD-RAM, etc. Windows 95/98/NT. $400.
  - *DVD Cut Machine*. Hardware audio/video encoder bundled with DVD Toolbox software. $800.

Who Can Produce a DVD for Me?

There are various steps to producing a DVD, but they can be split into two major parts: 1) *authoring* (creating the content and formatting a disc image), and 2) *replication* (cutting a master disc and stamping out hundreds or millions of copies). See the beginning of this chapter for more details.

[A] = authoring (including encoding, DVD-R duplication, and premastering).

[R] = replication (mastering, check discs, and mass production). Note that almost all replicators also have in-house authoring facilities or partnerships with authoring houses.

Other lists are available at DVDInsider, DVDMadeEasy, and Post Magazine. Also see "How Do I Copy My Home Videos/Movies/Slides to DVD?" for companies specializing in video-to-DVD-R transfers.

- [A] 12CM Multimedia (www.12cm.com, Mountain View, CA, 650-564-9000; Santa Clara, CA 408-350-9000).
- [A] 24-7DVD (dvd@24-7dvd.co.uk, Mogger Hanger, UK), +44 (0) 7764 187388.

- [A] Abbey Road Interactive (www.abbeyroad.co.uk, London, UK), +44 171 266 7000.

- [A] Acutrack (www.acutrack.com, Pleasanton, CA), 888-234-3472.

- [A] Advanced Media Post (www.ampost.com, Burbank, CA), 818-973-1668.

- [A] Advanced Visual Communications (AVCOM@aol.com, AVCOM Video) (Tampa, FL), 813-875-0888.

- [A] All Post (CA), 818-556-5756.

- [R] Americ Disc [also see MPO], www.amerdisc.com, Salida, CA, 888-545-7350; Miami, FL, 800-364-0759; Drummonville, Quebec, Canada, 800 263-0419.

- [A]Ascent Media (www.ascentmedia.com, Santa Monica, CA, 818-840-7235; Northvale, NJ, 201-784-2129). (Merger of companies including 4MC, ToddAO, and POP)

- [A] asv multimedia (asvid@gmx.de, Mengen, Germany), +49 (0) 7572-78361.

- [A] Atelier Digital (www.atelierdigital.com, Birmingham, AL), 205-263-7678.

- [A] Audio Plus Video International, www.apvi.com, Northvale, NJ, 201-767-3800; Burbank, CA, 818-841-7100.

- [A] AVCA (www.avca.com, Austin, TX), 512 472-4995.

- [A] AVM Dialog AB (www.avmdialog.se, Goteborg, Sweden).

- [A] B1 Media (www.b1media.com, Sherman Oaks, CA), 818-905-9902.

- [A] BCD Associates (www.bcdusa.com, Oklahoma City, OK), 405-843-4574.

- [A] Blink Digital (www.blinkdigital.com, New York, NY), 212-661-6900.

- [A] Blue City Digital (dvd@rbw.net, North Kansas City, MO), 816-300-0441.

- [A] California DVD (www.caldvd.com, San Francisco, CA), 1-800-864-1957.

- [A] Cambridge Multimedia (www.cmgroup.co.uk, Cambridge, UK), +44 (0) 1954 262030.

- [A] CAT Technologies (www.thecatwebsite.com, London, UK), +44 (0)20 8332 6548.

- [A] CDA (www.cda.de/index.php4, Albrechts, Germany), +49 (0) 36 81 / 3 87 - 1 53.

- [R] CD Digital Card (www.cddigitalcard.com, Rancho Cucamonga, CA), 800-268-1256 [specialize in shaped discs].

- [R] CDman (www.cdman.com, Vancouver, BC, Canada), 800-557-3347.

- [R] CD Press (www.cd-press.ch, Bergdietikon, Switzerland), +41 (0)1 745 90 60.

- [R] CD-ROM-Works (www.cd-rom-works.com, Portland, OR), 503-219-9331.

- [A] Chicago Recording Company (Chicago, IL), 312-822-9333.

- [R] Cine Magnetics (www.cinemagnetics.com), Armonk, NY, 914-273-7500; Studio City, CA, 818-623-2560), 800-431-1102.

- [A] Cinram/POP DVD Center (www.dvdinfo.com, Santa Monica, CA).

- [R] Cinram (www.cinram.com, Huntsville, AL, 256-859-9042; Anaheim, CA, 714-630-6700; Richmond, IN, 800-865-2200; Scarborough, Ontario, Canada, 416-298-8190), 800-433-DISC.

- [A] CKS|Pictures (ianni@cks.com, CA & NY), 408-342-5009.

- [A] ComChoice (www.comchoice.com, Gardena, CA), 877-633-4241.

- [A] Complete Post (www.completepost.com, Hollywood, CA), 323-860-7622.

- [R] Concord Disc Manufacturing (www.concorddisc.com, Anaheim, CA), 714-666-2266.

- [A] Crafted Timbre (http://home.twcyny.rr.com/craftedtimbre, Cortland, NY) 607-756-4780.

- [R] Crest National (www.crestnational.com, Hollywood, CA), 323-860-1300.

- [A] CruSh Interactive, (www.crushinteractive.com, Houston, TX), 713-972-1133.

- [A] Cubist Post & Effects (www.cubistpost.com, Philadelphia, PA), 215-627-1292.

- [A] CustomFlix (www.customflix.com, San Luis Obispo, CA), 978-626-1110.

- [A] Cut & Copy (studiowien@cutcopy.com, Vienna, Austria), +43 1 523 98 24.

- [A] CVC (www.cvcdigital.com, Los Angeles, CA), 818-972-0200. (Time Warner California Video Center)

- [A] D2 Productions (www.d2prod.com, CA), 818-576-8113.

- [A] Dallas Digital Transfer (www.dallasdigitaltransfer.com, Dallas, TX), 214-336-6292.

- [R] Davenport (Van Nuys, CA).

- [A] DAVID (www.davidmedia.it, Aprilia, Italy), 39-6-92704597.

- [R] Deluxe Video Services (www.bydeluxe.com, Carson City, CA), 310-518-0710. (Formerly Pioneer Video Manufacturing)

- [R] Denon Digital (now MD Digital)

- [A] Designlab Systems, (www.designlab.co.uk, London, UK), +44 (0) 207 437 5621.

- [A] Digidisc (www.digi-disc.com, Atlanta, GA), 770-925-1839.

- [A] Digisonics DVD (www.digisonics.com, Northridge, CA), 818-882-3444.

- [A] Digital Farm (www.digitalfarm.com, Seattle, WA), 206-634-2677.

- [A] Digital Group (London, UK)

- [A] digital images (www.digim.de, Halle, Germany), +49 (0)345/2175 -101.

- [A] Digital Metropolis (www.digitalmetropolis.com, Denver, CO), 303-292-4692.

- [A] Digital Outpost (www.dop.com, CA), 800-464-6434.

- [A] Digital Safari (www.digitalsafari.net, UK), +44 (0)7092 144 480.

- [A] Digital Video Compression Corporation (www.dvcc.com, CA), 818-777-5185.

- [A] Digital Video Dynamix (Seaford, NY), 516-826-6414.

- [A] Digital Video Mastering (Sydney, Australia).

- [A] Digitonium (www.digitonium.com, Los Angeles, CA), 818-889-2215.

- [A] Digiverse (www.digiverse.co.uk, London, UK), +44 (0) 20 7287 3141.

- [R] DISC (Orem, UT).

- [R] Disc Makers (www.discmakers.com, Pennsauken, NJ; Fremont, CA), 800-468-9353.

- [R] Disc Manufacturing Inc. (now part of Cinram).

- [R] DiscBurn.Com (St. Paul, MN), 612-782-8200.

- [R] Disctronics (www.disctronics.com, Southwater, UK; Plano, TX; Saint Mande, France; Italy).

- [A] Directorsite (www.directorsite.net, Manhattan Beach, CA), 310-727-2770.

- [A] DGP (www.dgpsoho.com, London, UK), +44 0 207 734 4501.

- [R] DOCdata (www.docdata.com, Tilburg, The Netherlands, +31 13 544 6444; Berlin, Germany, +49 30 467 0840; Sanford, ME, USA, 207-324-1124; Canoga Park, CA, USA 818-341-1124).

- [A] DownStream Digital (www.downstream.com, Portland, OR), 503-226-1944.

- [A] DVD Austin (www.dvdaustin.com, Round Rock, TX), 800-831-3774.

- [A] DVD Labs (www.dvdlabs.com, Princeton, NJ), 888-DVD-LABS.

- [A] DVD Master (www.dvdmaster.com, Fountain Valley, CA), 714-962-4098.

- [A] DVD Power (www.dvdpower.co.nz, Auckland, New Zealand), +64 (9) 415 5639.

- [A] DVD Power (www.dvdpwr.com, Singapore), +65 7796155.

- [A] DVD Recording Center (www.dvd-recording.com, Acton, MA), 800-321-8141.

- [A] DVD Technologies (www.dvdtech.com.au, Sydney, Australia), 1-300-FOR-DVD.

- [A] DVD Transfer.com (www.dvdtransfer.com, Minneapolis, MN), 612-676-1165.

- [A] DVDworx (www.dvdworx.com, Philadelphia, PA), 215-238-9679.

- [A] Dynamic Media (www.dynamicmediainc.com, Ellicott City, MD), 410-203-2553.

- [R] DV Line (e-yu882@hitel.net, Seoul, Korea), 82-2-3462-0331.

- [A] DVM - Digital Video Mastering (www.dvm.com.au, Sydney, Australia), +61 2 9571 6767.

- [A] EagleVision (www.eaglevisiontv.com, Stamford, CT), 800-EAGLE73.

- [R] Ecofina (www.ecofina.it, Milan, Italy), +39 024816121.

- [A] EDS Digital Studios (CA), 213-850-1165.

- [A] Electric Switch (lindsayh@owl.co.uk, London), +44-0-131-555-6055.

- [R] EMI Operations Italy (www.emioperations.it, Caronno Pertusella (VA), Italy), +39 02 965111.

- [A] E-M-S (www.e-m-s.de, Dortmund, Germany), 0231 442411-0.

- [A] Ent/Gates Productions (www.entgates.com, Buffalo, NY), 716-692-0064.

- [A] escape lab (www.escapelab.com, Brussels, Belgium) +32 2 644 99 62.

- [R] Euro Digital Disc (www.euro-digital-disc.de, Görlitz, Germany), +49 (0) 35 81 - 85 32 0.

- [A] FatDisc (www.fatdisc.com, Seattle, WA), 206-547-3055.

- [A] Film- und Videotechnik B. Gurtler (Munchen, Germany).

- [A] Firefly (Ireland).

- [A] Fitz.com (Santa Monica, CA) 310-315-9160.

- [A] Flare DVD (London, UK), +44 (0) 20 7343 6565

- [A] Forest Post Productions (www.ForestPost.com, Farmington Hills, MI), 248-855-4333.

- [A] Full Circle Studios (www.fullcirclestudios.com, Buffalo, NY), 716-875-7740.

- [A] FULLSTREAM DVD (www.fullstreamdvd.com, Dallas, TX), 214-969-1820.

- [R] Future Media Productions (www.fmpi.com, Valencia, CA), 661-294-5575.

- [A] Future Disc Systems (www.futurediscsystems.com, West Hollywood, CA), 323-876-8733.

- [A] G9 Interactive (www.gatewaymastering.com, Monrovia, CA), 626-358-0859.

- [A] Gateway Mastering Studios (Portland, ME), 207-828-9400.

- [R] Gema OD (Madrid, Spain), +34 91 643 42 55,

- [A] Gnome Digital Media (www.gnomedigital.com, Burbank, CA), 818-563-6539.

- [R] GoldenROM (www.goldenrom.com, Canonsburg, PA), 888-757-3472.

- [A] GTN (www.gtninc.com, Oak Park, MI), 248-548-2500.

- [A] GVI (www.g-v-i.com, Washington, DC), 202-293-4488.

- [A] HAVE (www.haveinc.com, Hudson, NY), 518-828-2000.

- [A] hdmg (www.hdmg.com, Minneapolis, MN), 952-943-1711.

- [A] HD Studios [DVD-Audio only] (www.hd-studios.com, CEDEX, Suresnes, France).

- [A] Hecker & Schneider GmbH (Dortmund, Germany).

- [A] Henninger Interactive Media (www.dvdexperts.com, Arlington, VA), 703-243-3444.

- [A] HNC Video/DVD Production (chip@HouseNation.com, Chicago, IL), 847-338-6560.

- [A] Hoboken Entertainment (Los Angeles, CA), 310-470-9440.

- [A] Hoek & Sonépouse (www.hoek.nl, Diemen, The Netherlands), +31 020 - 69 09 141.

- [R] Home Run Software Services (www.home-run.com, Huntington Beach, CA), 714-375-5454.

- [A] Ibis Multimedia (www.ibismultimedia.co.uk, Suffolk, UK), +44 01473 288865.

- [A] IBM InteractiveMedia (interactive@vnet.ibm.com, GA), 770-835-7193.

- [A] IBT Media (www.ibtmedia.com, Merriam, KS), 913-677-6655.

- [R] Imation (formerly 3M) (WI), 612-704-4898.

- [A] Immediate Impact (www.d-v-d.net, UK), +44 01322 553 505.

- [R] Infodisc (www.infodisc.com.tw, Taipei, Taiwan, 886-2-22266616; El Paso, TX).

- [A] Instinct Video & Film Productions (www.instinctfilms.com, Orlando, FL), 407-647-9555.

- [A] International Digital Centre (IDC) (www.idcdigital.com, New York, NY), 212-581-3940.

- [A] IPA Intermedia (www.ipahome.com, IL), 773-871-6033.

- [R] IPC Communication Services (www.ipc-world.com, Foothill Ranch, CA), 949-588-7765.
- [A] JamSync (www.jamsync.com, Nashville, TN), 615-320-5050.
- [A] Javanni Digital Video (www.javanni.com, Atlanta, GA), 704-795-7712.
- [R] JVC Disc America (www.jvcdiscusa.com, Sacramento, CA), 310-274-2221.
- [R] KAO Infosystems (www.kaoinfo.com, Fremont, CA), 800-525-6575.
- [R] Kao (Ontario, Canada), 800-871-MPEG.
- [A] kdg mediatech (www.kdg-mt.com, Elbigenalp, Austria, +43 (0) 5634-500; Parc d'Activités, France, +33 (0) 3 29 58 40 70).
- [A] k-kontor[Hamburg] kommunikations (www.k-kontor.de, Hamburg, Germany), +49-40-850-9021.
- [A] The Lawrence Company (www.lawrenceco.com, Santa Monica, CA), 310-452-9657.
- [R] LaserPacific (www.laserpacific.com, CA), 213-462-6266.
- [R] Lena Optical Disc (www.lenaoptical.com, Hong Kong), 852-2556-8198
- [A] Look and Feel New Media (www.lookandfeel.com, Kansas City, MO), 816-472-7878.
- [A] The Machine Room (www.themachineroom.co.uk, London, UK), +44 171 734 3433.
- [A] Mares Multimedia (www.maresmultimedia.com, Nashville, TN), 615-356-3905.
- [A] Marin Digital (www.marin-digital.com, Sausalito, CA), 415-331-4423.
- [A] Main Point Interactive (www.mainpoint.com, Oley, PA), 610-987-9320.
- [R] Marcorp (www.marcopr.com, Pittsburgh, PA), 800-284-6277.
- [A] Mastering Studio München (www.msn-muenchen.de, Munich, Germany), +49-89-286692-0.
- [R] MD Digital Manufacturing (Madison, GA), 706-342-3425.
- [R] Maxell Multimedia (now MD Digital).
- [R] Maxwell Productions (Scottsdale, AZ).
- [A] Media Tech (www.mediatech1.com, Denver, CO), 303-741-6878.

- [R] Memory-Tech Corporation (Tokyo, Japan).

- [A] MEP Medienhaus (www.mep-ffm.com, Frankfurt, Germany), +49 (0)69 78960202.

- [R] Mercury Entertainment (www.jaring.my/mercury, Selangor Darul Ehsan, Malaysia).

- [R] Metatec (www.metatec.com, Dublin, OH, 614-761-2000; Milpitas, CA, 408-519-5000; Breda, www.metatec.nl, Netherlands, +31 76 5333 100)

- [A] Metropolis Group (www.metropolis-group.co.uk, London, UK), +44-20-8742-1111.

- [A] Microsoft Studios Digital Video Services (dvs@microsoft.com, Redmond, WA).

- [A] Microvision Services (www.microvision.org.uk, Huddersfield, UK), +44 1484 644852.

- [A] Mills/James Productions (www.milljames.com, Columbus, OH), 614-777-9933.

- [A] Mirage Video Productions (www.miragevideo.com, Boulder, CO), 303-786-7800.

- [A] MPEG Production (www.mpeg.se/index.htm, Stockholm, Sweden), +46-8-324030.

- [R] MPO (www.mpo.fr, Europe, North America, and Asia), +33 01 41 10 51 51.

- [R] MRT Technology [Ritek partner] (www.mrttech.com, City of Industry, CA), 626-839-5555.

- [R] Multimedia Info-Tech [Ritek partner] (www.mitcds.com, Belfast, Ireland), +44 (0) 2890 300883.

- [R] Multi Media Replication (www.replication.com, Andover, UK) +44 (0)1264 336 330.

- [R] Nimbus CD International (www.nimbuscd.com, see Technicolor).

- [A] NOB Interactive (www.nob.nl, Netherlands), +31 (0)35-677-5413.

- [A] NordArt Video & Multimedia (www.nordart.se, Sundbyberg, Sweden), +46 8764 66 90.

- [R] Nordisc (www.nordisc.no, Rjukan, Norway), +47 35 08 01 00.

- [A] Oasis Post (www.oasispost.com.au, Kent Town, South Australia), +61 8 8362 2888.

- [A] Oasis Television (www.oasistv.co.uk, London, UK), +44 (0) 20 7534 1808.

- [R] OEM (www.oemdisc.com, Charlotte, NC), 704-504-1877.

- [R] Optical Disc Corporation, 562-946-3050. (www.optical-disc.com, LaserWave DirectCut DVD recorder for creating single copies.)

- [R] Optical Disc Media (CA).

- [R] Optimes (www.optimes.it, L'aquila, Italy), +39-0862-3311.

- [A] Option Facilities (www.option-av.be, Mechelen, Belgium), +32/15/28 73 00.

- [A] The Other Side [nee TwoPlusOne] (www.twoplusone.co.uk, London, UK), +44 (0) 207 494 8290.

- [A] OUTPOST (merrickgroup@earthlink.net, Charlotte, NC), 704-344-3577.

- [A] Pacific Coast Sound Works (CA), 213-655-4771.

- [R] Pacific Mirror Image (Melbourne, Australia).

- [A] Pacific Ocean Post (CA), 310-458-9192. (Now part of Ascent Media)

- [A] Pacific Video Resources (www.pvr.com, CA), 415-864-5679.

- [R] Panasonic Disc Services Corp (www.panasonicdvd.com, Torrance, CA; Pinckneyville, IL; Guadalajara, Mexico; Youghal, Ireland), 310-783-4800.

- [A] Paris Media System (Paris, France).

- [A] Paul Stubblebine Mastering and DVD (www.paulstubblebine.com, San Francisco, CA), 415-469-0123

- [A] The Pavement (www.the-pavement.com, London, UK), +44 (0) 207 426 5190.

- [A] Performance Digital Labs (www.performancedigital.com, San Diego, CA), 800-253-3085.

- [A] Phaebus (www.phaebus.co.uk, Manchester, UK), +44 (0) 161 950 8105.

- [A] PIMC (Professional Interactive Media Centre) (www.pimc.be, Diepenbeek, Belgium), +32 11 303690.

- [A] Pioneer France (Nanterre, France), 33 1 47 60 79 30.

- [R] Pioneer Optical Disc (Barcelona, Spain), +34-93-739-99-00.

- [R] Pioneer Video (www.pvc.co.jp, Kofu, Japan).
- [A] Pixel Farm Interactive (www.pixelfarm.com, Minneapolis, MN), 612-339-7644.
- [A] Positive Charge Ltd. (www.positivecharge.com.pl, Warszawa, Poland), +48 22 632 97 32.
- [R] Pozzoli (www.pozzolispa.com, Milan Italy) +39 02 954341.
- [A] PRC Digital Media (www.prcdigital.com, Jacksonville, FL), 904-354-5353.
- [A] Prime Disc [Ritek partner] (www.primedisc.com, Wiesbaden, Germany), +49-611-9628644.
- [A] Provac Disc Media (anne.brown@sympatico.ca, Toronto, Ontario), 800-876-9013.
- [R] Racman Avdio Video Studio (www.racman.si, Ljubljana, Slovenia), +386 1 5819-201.
- [A] Riccelli Creative (www.riccellicreative.com, Fort Worth, TX), 817-332-7777.
- [A] The Richard Diercks Company (www.dvdauthor.com, Minneapolis, MN), 612-334-5900.
- [R] Ritek (www.ritek.com, HsinChu, Taiwan, ROC, +886-3-598 5696; Taipei, Taiwan, ROC, +886-2-8521-5555). [Also see MRT U.S.), Multimedia Info-Tech (Ireland), Prime Disc (Germany), and Ritek Australia.]
- [R] Ritek Australia (www.ritek.com, Alexandria, Australia), +61-2-9669-3311).
- [A] Rivetal (www.rivetal.com, Provo, UT), 801.818.2222.
- [R] Saturn Solutions (www.saturndisc.com, Markham, Ontario, 905-470-0844; St. Laurent, Quebec, 514-856-5656; Provo, Utah, 801-370-9090; Dublin, Ireland, +353-1-403-8599).
- [A] ScreamDVD (www.screamdvd.com, New York, NY), 212-951-7171.
- [R] SDC Group (www.sdc-group.com, Brabrand, Denmark), + 45 87 45 45 45.
- [A] Sharpline Arts (www.sharplinearts.com, Glendale, CA), 818-500-3958.
- [R] SKC (www.sk.com, Chonan, South Korea).
- [R] SNA (www.snacompactdisc.com, Tourouvre, France), +33 (0) 2 33 85 15 15.

- [R] Sonopress (www.sonopress.de, Gütersloh, Germany), +49-5241-80 5200; Weaverville, NC, USA, 828-658-2000; Dublin, Ireland, +353 1 840 9000; Madrid, Spain, +34-91-6 71 22 00; Forbach, France, +33-1-53 43 82 32.

- [R] Sony DADC (www.sonydadc.com, Niederalm, Austria), +43 624 688 0555.

- [R] Sony Disc Manufacturing (http://sdm.sony.com, Terre Haute, Indiana), 800-358-7316.

- [A] Sound Chamber Mastering (www.sound-chamber.com, North Hollywood, CA), 818-752-7581.

- [A] SOUNDnVISION (m.fiocchi@snv.it, Milano, Italy), +39 02 55 18 02 45.

- [R] Spool Multi Media (www.smmuk.co.uk, Deeside, UK), +44 (0) 1244 280602.

- [A] Squash DVD (www.squashpost.co.uk, London, UK), +44 (0) 20 7292 0222.

- [A] Star Video Duplicating (www.starvideo.com, Phoenix, AZ), 602-437-0646.

- [A] Stay Tuned (www.staytuned.be, Brussels, Belgium), +32 2 7611100.

- [A] Stimulus (Calgary, Alberta).

- [A] Sté EXILOG (EXILOG.JML@WANADOO.FR, Vendoeuvres FRANCE), 33 02 54 38 30 95.

- [A] Stonehenge Filmworks (www.stonehenge.ca, Toronto and Ontario, Canada), 416-867-1189.

- [A] Stream AV (www.streamav.com.au, Melbourne, Australia), +61 3 9376 6444.

- [A] Studio Reload (www.reload.tv, Boise, ID), 208-344-4321.

- [A] Sunset Post (www.sunsetpost.com, CA), 818-956-7912.

- [A] Super Digital Media (www.superdvd.com, Santa Clara, CA), 408-727-5091.

- [A] Sync Sound (NY), 212-246-5580 (5.1 audio).

- [A] Syrinx music & media GmbH (www.syrinx.de, Hamburg, Germany), +49-40-63709230.

- [A] Systeam (www.systream.it, Rome, Italy), +39-06-508141.

- [A] Tape House Broadband (www.tapehousebroadband.com, New York, NY), 212-557-4949.

- [R] Takt (www.takt.pl, Warsaw, Poland), +48 22 874 35 75.

- [A] TC Video (tcvideo@btinternet.com, Middlesex, UK), +44 (0)208 904 6271.

- [R] Technicolor (scorso@nimbuscd.com, Camarillo, CA, 805-445-1122; Charlottesville, VA, 804-985-1100; Cwmbran, Wales, UK, 44-1163-465-000), 800-732-4555).

- [R] TIB (www.tib.co.uk, Merthyr Tydfil, UK), + 44 (0)1685 354700.

- [A] Tobin Productions (www.tobinproductions.com, New York, NY), 212-727-1500.

- [R] Tocano (www.tocano.dk, Smoerum, Denmark), +45 44666200.

- [A] Tree Falls (www.tfsound.com, Los Angeles, CA), (323) 469-1068.

- [R] Universal Manufacturing & Logistics (www.u-m-l.com, Blackburn, UK, +44 (0) 1254 505300; Langenhagen, Germany, +49 (0) 511-972-1755).

- [A] Valkieser Solutions (www.valkieser.nl, Hilversum, Netherlands), +31-35-6714-300.

- [R] Japan Victor (Kanagawa, Japan), 45-453-0305.

- [A] Video Movie Magic (www.videomoviemagic.com, Laguna Hills, CA), 949-582-8596.

- [A] Video Replay (www.videoreplaychicago.com, Chicago, IL),

- [A] Video Transfer (www.vtiboston.com, Boston, MA), 617-247-0100.

- [A] Visible Light Digital (www.vldigital.com, Orlando, FL), 407-327-7804.

- [A] Visom Digital (www.visomdigital.com.br, Rio de Janeiro, Brazil), +55 21 539-7313.

- [A] The Vision Factory (www.tvf.com, St. Louis, MO), 314-963-7887.

- [A] Vision Wise (www.visionwise.com, Irving, TX), 888-979-9473.

- [R] Warner Advanced Media Operations (www.ivyhill-wms.com, WAMO), 717-383-3291.

- [A] The Zak Studio (www.thezakstudio.com, Paris, France), +33 1 49823773.

- [R] Zomax, (www.zomax.com, Plymouth, MN, 612-577-3515; Fremont, CA, 510-492-5191; Indianapolis, IN, 510-492-5191; Dublin, Ireland, 353-1-405-6222; Langen, Germany, 49-6103-9702-23).

- [A] Zuma Digital. Now part of Tape House Broadband (www.tapehousebroadband.com).

What Testing/Verification Services and Tools Are Available?

- AudioDev (www.audiodev.com, Sweden, USA, Hong Kong), +46 40 690 49 00.

- CD Associates (www.cdassociates.com, CA). Testing equipment and software. (714) 733-8580.

- ContentWise (www.contentwise.com, Rehovot, Israel), +972-8-940-8773. *Second Sight* software for checking compatibility of DVD titles on multiple players.

- Hitachi (Japan). Testing services and test discs. Official DVD Forum verification lab.

- Intellikey Labs (www.intellikeylabs.com, Burbank , CA), (818) 953-9116, fax (818) 953-9144.

- Interra Digital Video Technologies (www.interra.tv/), *Surveyor* software, $6,000. *DProbe*, $10,000.

- ITRI (www.oes.itri.org.tw, HsinChu, Taiwan). Testing services and test discs. Official DVD Verification Lab. 886-3-591-5066, fax 886-3-591-7531.

- Matsushita (Japan). Testing services, test discs, and test equipment. Official DVD Verification Lab. +81-6-6905-4195, fax +81-6-6909-5027.

- Matsushita/Panasonic (Japan). *Panasonic LQ-VD300P* emulator. Hardware player with Windows NT software. $15,000.

- Philips (www.philips.com, Europe), *DVD-Video Verifier* software, $500. Official DVD Verification Center.

- Pioneer (Japan). Testing services and test discs. Official DVD Verification Lab. +81-3-3495-5474, fax +81-3-3495-4301.

- PMTC (Professional Multimedia Test Centre) (www.pmtctest.com/new/upgrade_flash/upgrade_flash.html, Diepenbeek, Belgium), +32 11 303636.

- Sonic Solutions (www.sonic.com, USA). *DVD PrePlay* software. Emulation and diagnosis tools for Windows. $5000.

- Sony (Japan). Testing services and test discs. Official DVD Format Lab. +81-3-5448-2200, fax +81-3-5448-3061.

- Testronic Labs (www.testroniclaboratories.com/, Burbank, CA), (818) 845-3223,
  fax (818) 845-3236.

- Toshiba (Japan). Testing services and test discs. Official DVD Verification Lab. +81-3-3457-2105, fax +81-3-5444-9202.

- Victor (Japan). Testing services and test discs. Official DVD Verification Lab. +81-3-3289-2813, fax +81-45-450-1639.

- WAMO (www.ivyhill-wms.com, USA). Testing services and test discs. Official DVD Forum verification lab. 1-570-383-3568,
  fax 1-570-383-7487.

Also see "Other Production Tools" for tools to analyze and verify coded bitstreams, disc images, and DLTs.

Can I Put DVD-Video Content on a CD-R or CD-RW?

NOTE: This section refers to creating original DVD-Video content, not copying from DVD to CD. The latter is impractical, since it takes 7 to 14 CDs to hold one side of a DVD. Also, most DVD movies are encrypted so that the files can't be copied without special software.

There are many advantages to creating a DVD-Video volume using an inexpensive recordable CD rather than an expensive recordable DVD. The resulting "cDVD" (also called a "miniDVD") is perfect for testing and for short video programs. Unfortunately, you can put DVD-Video files on CD-R or CD-RW media, or even on pressed CD-ROM media, but almost no set-top player can play the disc. There are a number of reasons DVD-Video players can't play DVD-Video content from CD media:

1. Checking for CD media is a fallback case after DVD checking for, at which point the players are no longer looking for DVD-Video content

2. It's simpler and cheaper for players to spin CDs at 1x speed rather than the 9x speed required for DVD-Video content

3. Many players can't read CD-R discs (see "Is CD-R compatible with DVD?" in Chapter 2, "DVD's Relationship to Other Products and Technologies").

The only known players that can play a cDVD are the Afreey/Sampo LD2060 and ADV2360 models, and the Aiwa XD-DW5 and XD-DW1. Some of these players use 1x or 2x readers so they can't handle data rates over 4 Mbps. It's possible to replace the IDE drive mechanism in the player with a faster drive, which can then handle higher data rates. See robshot.com for details on cDVD-capable players. (Note: there have been many reports of players able to play DVD content from CD-R. Upon investigation it turns out that they play Video CDs but not cDVDs. The players mentioned above have been verified to play DVD-Video files (.VOB and .IFO) from CD media.)

Computers are more forgiving. DVD-Video files from any source with fast enough data rates, including CD-R or CD-RW, with or without UDF formatting, will play on most DVD-ROM PCs as long as the drive can read the media (all but early model DVD-ROM drives can read CD-Rs). On a Mac, you need version 2.3 or newer of the Apple DVD Player.

To create a cDVD, author the DVD-Video content as usual (see "What DVD Authoring Systems Are Available?") then burn it to a CD-R or CD-RW. If your authoring software doesn't write directly to CD-R/RW discs, use a separate utility to copy the VIDEO_TS directory to the root directory of the disc. To be compatible with the few settop players that read cDVDs, turn on the UDF filesystem option of the CD burning software. To achieve longer playing times, encode the video in MPEG-2 half-D1 format (352 × 480 or 352 × 576) or in MPEG-1 format.

An alternative is to put Video CD or Super Video CD content on CD-R or CD-RW media for playback in a DVD player. Set-top DVD players that are VCD or SVCD capable and can read recordable media will be able to play such discs (see "Is Video CD Compatible with DVD?" in Chapter 2). The limitations of VCD apply (MPEG-1 video and audio, 1.152 Mbps, and 74 minutes of playing time). All DVD-ROM PCs able to read recordable CD media can play recorded VCD discs. An MPEG-2 decoder (see "Can I Play DVD Movies on My Computer?" in Chapter 4) is needed to play SVCDs. See "How Do I Copy My Home Videos/Movies/Slides to DVD?" for more on creating Video CDs.

How Do I Copy My Home Video/Film/Photos to DVD?

This used to be almost impossible, but luckily for you it's getting cheaper and easier all the time.

For a simple video-to-DVD transfer you can buy a DVD video recorder ($500 to $3,000) and connect it to your VCR or camcorder. It works just like a VCR but it records onto a disc instead of tape.

For transferring photos, or for making a customized DVD with menus, chapters, and other fun stuff, you'll need the following:

- A computer
- A DVD recordable drive ($200 to $600, or it might come with the computer)
- DVD authoring software (usually comes with the drive or computer, or you can buy it for $40 to $27,000, see "What DVD Authoring Systems Are Available?")

NOTE: You must use authoring software. You can't just put MPEG or AVI files on a disc and expect it to play in DVD players.

Then take the following steps

- If the video and pictures are not already in digital form (AVI, WMV, DivX, QuickTime, JPEG, and so on) you need to transfer them to your computer. For analog video, such as VHS and Hi8, you need a video capture device or a computer with built-in analog video input; for digital video such as DV or D8 you need a 1394/FireWire input on the computer. For film, first have it transferred to tape or digital video at a camera shop or video company. For slides or photos, use a scanner (or rent scanning time at a place such as Kinkos).
- Import the video and audio clips into the DVD-Video authoring program. Many DVD authoring programs will convert and encode the video and audio for you. If not, you have to
  - Encode the video into MPEG-2 (make sure the display frame rate is set to 29.97 for NTSC or 25 for PAL).
  - Encode the audio into Dolby Digital (or, if your video is short enough that you have room on the disc, format the audio as 48 kHz PCM). You can also use MPEG Level 2 audio, but it won't work on all players.
- Create some chapter points in your video tracks or let the DVD recording software do it for you.
- To put photos on the disc, use the slideshow feature in the authoring software or make each picture a menu. Most DVD authoring software will directly read TIFF, JPEG, BMP, and Photoshop files.
- Create menus that link to your video clips and slideshows.
- Write your finished gem out to a recordable DVD ($2-$15). (But see "Is It True There Are Compatibility Problems with Recordable DVD Formats?" in Chapter 4 for compatibility worries.)

John Beale has written a page about his experiences making DVDs (www.bealecorner.com/trv900/DVD/authoring.html).

Another option is to use a service that does all the work for you at a reasonable fee. Here are a few choices.

- 3-Lib (http://3lib.ukonline.co.uk/dvd/, Reading, UK). Up to 2 hours for £25. PAL format.

- American Digital Media (www.AmericanDigitalMedia.com, Hoover, AL). Up to 2 hours for $99.

- Digital Video Dynamics (www.vcdtransfer.com, Orlando, FL). Up to 2 hours for $40 (chapters at 5-minute intervals).

- DVD ELF (www.movieblowout.com, Miami, FL). Up to 1 hour for $60. 2 hours for $95.

- DVD Wedding Productions (www.dvdweddings.com, South Pasadena, CA). One tape for $150 (+ VHS dubbing charge).

- HomeMovie.com (Everett, WA). Up to 2 hours for $50 (chapters included).

- ImageStation (www.imagestation.com/a2dvd/, Sony/Vingage; Reston, VA). Up to 90 minutes for $40.

- Latale Productions (www.thevideodisc.com, Flushing, NY). 1 tape for $99 (chapters extra).

- ScreamDVD (www.screamdvd.com, New York, NY). Up to 1 hour for $40, up to 2 hours for $70 (chapters at 3-minute intervals).

- VHS-to-DVD (www.vhs-to-dvd.com, Pembroke Pines, FL). Up to 1 hour for $18 to $25, up to 2 hours for $28 to $35.

- Visualisation Systems (www.visualisationsystems.co.uk, Preston, UK). Up to 1 hour for £35. Up to 2 hours for £40. VCD for £20.

- Wedding DVD (www.weddingdvds.com) no longer offers the service.

- YesVideo.com (San Jose, CA; kiosks at Target, Walgreens, and elsewhere). $37 for 1 hour, $60 for 2 hours (chapters included).

Or, if near-VHS quality is sufficient, make a Video CD. Get MPEG-1 video encoding software and a CD-R/RW formatting application that supports Video CD such as *Easy CD Creator* or *Toast* from Roxio (formerly Adaptec), *InstantCD* from Pinnacle (formerly from VOB), *InternetDiscWriter* from Query, *MPEG Maker-2* from VITEC, *MyDVD* or *RecordNow Max* from Sonic, *Nero Burning ROM* from Ahead, *NTI CD-Maker* from NTI, or *WinOnCD* from Cequadrat. Quality is not as good as DVD, and playing time is not as long, but hardware and blank CDs are cheaper. Just make sure that any players you intend to play the disc in can read CD-Rs (see "Is CD-R compatible with DVD?" in Chapter 2) and can play Video CDs (see "Is Video CD compatible with DVD?" in Chapter 2). See VCDhelp.com for more on making Video CDs. A variation on this strategy is to make Super Video CDs (see "Is Super Video CD compatible with DVD?" in Chapter 2), which have better quality but shorter playing time. A few of the authoring/formatting tools listed above can make SVCDs, but few DVD players can play SVCDs.

Another option is a home Video CD recorder, such as the Terapin *CD Audio/Video Recorder* or the TV One *MPEG-2@disk*, which record video from analog inputs to CD-R or CD-RW.

How Can I Copy a DVD?

This section is about copying disc-to-disc. See "How Can I Record from DVD to Videotape?" in Chapter 2 for copying to tape.

First, please understand that copying a commercial DVD may be illegal, depending on what you do with the copy. Copying video for your own personal use is legal, but making copies of copyrighted discs for friends is not.

Second, be aware that almost all DVD movies are protected from casual copying. See "What Are the Copy Protection Issues?" in Chapter 1, "General DVD," for details. However, any protection measure is usually broken, see "What is DeCSS?" in Chapter 4.

Third, realize that many movies come on dual-layer discs (DVD-9s), which can't be directly copied to recordable DVD since there are no dual-layer recordable discs, although you may be able to break up the content from on DVD-9 onto two recordable discs.

Fourth, understand that copying the files from a DVD to a recordable DVD often produces a disc that won't play in a set-top DVD player, since the files have to go in specific order and specific places on the disc. Some DVD writing software recognizes the files and places them correctly, but other software doesn't. In other words, you can't just copy the .IFO and .VOB files (see "What are .IFO, .VOB, and .AOB files? How Can I Play Them?" in Chapter 4).

If you have a legitimate need to copy a DVD, such as a disc you made yourself, there are a number of options. You can hook a DVD player to a set-top DVD video recorder. Some DVD authoring software (see "What DVD authoring systems are available?") can import video from an unprotected disc. There are computer software utilities you can use to extract video and audio from a disc, which you can then use to make a new disc. There are also software tools for copying entire discs. See "DVD Utilities and Region-free Information" in Chapter 6, "Miscellaneous," and "Other Production Tools" for tools; see "How Do I Copy My Home Videos/Movies/Slides to DVD?" for how to make your own DVDs.

Beware of e-mail and ads touting DVD copying software for sale. See "What's with Those 'Copy Any DVD' E-mails?" below.

What's with Those "Copy Any DVD" E-mails?

It's true you can copy any DVD movie. However the people selling DVD copying software conveniently don't mention the many free alternatives, nor do most of them mention that their applications only copy to CD-R/RW in Video CD format, which means the video quality is crummy and the copies don't play in about half the DVD players out there (see "Is CD-R Compatible with DVD?" and "Is Video CD compatible with DVD?" in Chapter 2). They also neglect to mention that copying movies from rental stores or from friends is illegal.

How Do I Get a Job Making DVDs?

Read this book through a few times. For extra credit read my book, *DVD Demystified*, and visit some of the DVD information sources listed in "Where Can I Get More Information About DVD?" in Chapter 6. Then attend a conference (see "How Do I Get a Job Making DVDs?") to learn more and to make contacts in the DVD industry. Take a few training courses (see "How Do I Get a Job Making DVDs?"). Consider joining the DVDA. If you can, volunteer to be an intern at a DVD production house (see "Who Can Produce a DVD for Me?").

Once you have a little experience, you'll be in great demand!

Where Can I Get DVD Training?

A variety of workshops and seminars on various DVD topics are presented at conferences such as DVD Pro, DVD Summit (Europe), or DVD Production.

Training companies offer DVD courses and "boot camps":

- adicomm (www.adicomm.com, Costa Mesa, CA)
- dvd.learn (www.dvdlearn.com, Denver, CO)
- Ex'pression Center for New Media (www.xnewmedia.com, Emeryville, CA)
- Gnome Digital Media (www.gnomedigital.com, Burbank, CA), maker of the *DVD 101* training/template discs
- I.N.C. Technologies (http://inc-tech.com, Glendale, CA), oriented towards amateur DVD users
- TFDVD.com (Havertown, PA), *DVD Studio Pro* training

- Seneschal (http://seneschal.net, San Francisco, CA)
- Texas State Technical College (http://waco.tstc.edu/dvd/, Waco, TX)
- Video Symphony (www.videosymphony.com, Burbank, CA)

Some schools offer full-term courses:

- Ngee Ann Polytechnic Digital Media Authoring Studio (www.np. edu.sg, Singapore)
- Seneca College (http://scaweb.senecac.on.ca, Toronto, CA)
- South Seas Film and Television School (www.southseas.co.nz, Auckland, New Zealand)

The major DVD authoring software companies offer training classes around the world, sometimes for free:

- Apple Computer (www.apple.com)
- Sonic Solutions (www.sonic.com)

How Can I Sell DVDs That I Made?

- Amazon zShops sales referrals. Your disc is listed on Amazon site, Amazon processes orders, you are responsible for producing, packaging, and shipping discs.
- CustomFlix duplication and e-commerce consignment. You give them a disc (or tape that they turn into a disc), and they handle order processing, copying onto DVD-Rs, labeling, packaging, and shipment. No minimum is required.
- Auction sites such as eBay, Amazon Auctions, Yahoo Auctions, uBid, and many others. The Site runs the auction, you are responsible for taking payment, producing, packaging, and shipping discs.

If you are looking for someone to deliver your titles to retailers, see "Studios, Video Publishers, and Distributors" in Chapter 6 for distributors.

How Do I Put a PowerPoint Presentation on DVD?

There's not yet a feature in PowerPoint to export directly to video on DVD, but you can convert a PowerPoint presentation to stills or video for import into a DVD authoring program (see "How Do I Copy My Home Videos/Movies/Slides to DVD?"). Recent versions of PowerPoint allow you to save your slides as graphic images (JPEG or PNG files) that can be imported into a DVD authoring program that supports slideshows. The advantage of using the slideshow feature is that you can have the DVD player pause indefinitely on each still until you press the Enter or Play key on the remote control. (Note: make sure the authoring software supports true slideshows with "infinite stills," since many programs just render slides as video.) The disadvantage of using stills is that you won't get animations and other fancy PowerPoint effects. Alternatively you can record the PowerPoint presentation as a video file (use a PowerPoint add-in or a motion screen capture program) and import the video file into the DVD authoring program. This preserves the full visual effect but locks you into the timing you used when recording the presentation.

Chapter 6

Miscellaneous

Who Invented DVD and Who Owns It?
Whom to Contact for Specifications and Licensing?

DVD is the work of many companies and many people. There were originally two competing proposals. The MMCD format was backed by Sony, Philips, and others. The SD format was backed by Toshiba, Matsushita, Time Warner, and others. A group of computer companies led by IBM insisted that the factions agree on a single standard. The combined DVD format was announced in September of 1995, avoiding a confusing and costly repeat of the VHS versus Betamax videotape battle or the quadraphonic sound battle of the 1970s.

No single company "owns" DVD. The official specification was developed by a consortium of ten companies: Hitachi, JVC, Matsushita, Mitsubishi, Philips, Pioneer, Sony, Thomson, Time Warner, and Toshiba. Representatives from many other companies also contributed in various working groups. In May 1997, the DVD Consortium was replaced by the DVD Forum, which is open to all companies, and as of 2003 had over 220 members. Time Warner originally trademarked the DVD logo, and has since assigned it to the DVD Format/Logo Licensing Corporation (DVD FLLC). The written term "DVD" is too common to be trademarked or owned. See "Who Is Making or Supporting DVD Products?" for a list of companies working with DVD.

The official DVD specification books are available after signing a nondisclosure agreement and paying a $5,000 fee. One book is included in the initial fee; additional books are $500 each. Manufacture of DVD products and use of the DVD logo for nonpromotional purposes requires additional format and logo licenses, for a one-time fee of $10,000 per format, minus $5,000 if you have already paid for the specification. (For example, a DVD-Video player manufacturer must license DVD-ROM and DVD-Video for $20,000, or $15,000 if they have the spec.) Contact DVD Format/Logo Licensing Corporation (DVD FLLC), Shiba Shimizu Building 5F, Shiba-daimon 2-3-11, Minato-ku, Tokyo 105-0012, tel: +81-3-5777-2881, fax: +81-3-5777-2882. Before April 14, 2000, logo/format licensing was administered by Toshiba.

ECMA International has developed international standards for DVD-ROM (part 1, the smallest part of the DVD spec), available for free download as ECMA-267 and ECMA-268 from www.ecma-international.org. ECMA has also standardized DVD-R in ECMA-279, DVD-RAM in ECMA-272 and ECMA-273, and DVD+RW as ECMA-274 (see "What About Recordable DVD: DVD-R, DVD-RAM, DVD-RW, DVD+RW, and DVD+R?" in Chapter 4, "DVDs and Computers"). Unfortunately, ECMA has the annoying habit of spelling "disc" wrong. Also confusing, if you're not from Europe, is ECMA's use of a comma instead of a period for the decimal point.

The specification for the UDF file system used by DVD is available from www.osta.org.

Many technical details of the DVD-Video format are available at the DVD-Video Information page (www.mpucoder.com/dvd/).

Any company making DVD products must license essential technology patents from a Philips/Pioneer/Sony pool (3.5 percent per player, minimum $5; additional $2.50 for Video CD compatibility; 5 cents per disc), a Hitachi/Matsushita/Mitsubishi/Time Warner/Toshiba/Victor pool (4 percent per player or drive, minimum $4; 4 percent per DVD decoder, minimum $1; 7.5 cents per disc) and from Thomson. Patent royalties may also be owed to Discovision Associates, which owns about 1300 optical disc patents (usually paid by the replicator).

The licensor of CSS encryption technology is DVD CCA (Copy Control Association), a non-profit trade association with offices at 225 B Cochrane Circle, Morgan Hill, CA. There is a $10,000 initial licensing fee, but no per-product royalties. Send license requests to css-license@lmicp.com, technical info requests to css-info@lmicp.com. Before December 15, 1999, CSS licensing was administered on an interim basis by Matsushita.

Macrovision licenses its analog antirecording technology to hardware makers. There is a $30,000 initial charge, with a $15,000 yearly renewal fee. The fees support certification of players to ensure widest compatibility with televisions. There are no royalty charges for player manufacturers. Macrovision charges a royalty to content publishers (approximately 4 to 10 cents per disc, compared to 2 to 5 cents for a VHS tape).

Dolby licenses Dolby Digital decoders for $0.26 per channel. Philips, on behalf of CCETT and IRT, also charges $0.20 per channel (maximum of $0.60 per player) for Dolby Digital patents, along with $0.003 per disc.

An MPEG-2 patent license is required from MPEG LA (MPEG Licensing Adminstrator). Cost is $2.50 for a DVD player or decoder card and 4 cents for each DVD disc, although there seems to be disagreement on whether content producers owe royalties for discs.

Many DVD players are also *Video CD* (VCD) players. Philips licenses the Video CD format and patents on behalf of themselves, Sony, JVC,

Matsushita, CNETT, and IRT for $25,000 initial payment plus royalties of 2.5 percent per player or $2.50 minimum.

Nissim claims 25 cents per player and 78/100 of a cent for parental management and other DVD-related patents.

Various licensing fees add up to over $20 in royalties for a $200 DVD player, and about $0.20 per disc. Disc royalties are paid by the replicator.

Who Is Making or Supporting DVD Products?

Consumer Electronics

- Afreey: DVD-Video players
- Aiwa: DVD-Audio and DVD-Video players
- Akai: DVD-Video players
- Alba: DVD-Video players
- Alpine: DVD car navigation/entertainment
- Altec Lansing: DVD audio technology
- Amitech: DVD-Video players
- Amoisonic: DVD-Video players
- Apex Digital: DVD-Video players (made by VDDV; info at www. nerd-out.com/apex and aenow.com/apex/)
- Arcam: DVD-Video players (UK)
- Ariston: DVD-Video players
- Atlantis Land: DVD-Video players
- A-trend: DVD-Video players
- Atta: DVD-Video players
- Audiologic: DVD-Video players
- Audiosonic: DVD-Video players
- Audiovox: Car DVD players
- Axion: DVD-Video players
- AV Phile (Raite): DVD-Video Players
- Bluesky: DVD-Video players
- BUSH: DVD-Video players
- California Audio Labs: DVD-Video players
- CAT: DVD-Video players
- Camelot: DVD-Video players
- Casio: DVD-Video players

- CCE: DVD-Video players
- CD Playright: protective film for discs
- Centrum: DVD-Video players
- Chunlan: DVD-Video players
- Clairtone: DVD-Video players
- Clarion: DVD car navigation/entertainment
- Comjet: DVD-Video players with Web connection
- Compro: DVD-Video players
- Conia: DVD-Video players (Australia, made by VDDV)
- Cougar: DVD-Video players
- Cyberhome (Yamakawa/Raite): DVD-Video players
- Daewoo Electronics: DVD-Video players
- Dantax: DVD-Video players
- Denon: DVD-Audio and DVD-Video players
- Denver: DVD-Video players
- Digitor: DVD-Video players
- Digitron: DVD-Video players
- DiViDo: DVD-Video players (Netherlands)
- Dual: DVD-Video players
- DVDO: video deinterlacing processors
- Dynamic: DVD-Video players
- Eagle Wireless International: DVD Internet appliances
- Eclipse: DVD-Video players
- Electrohome: DVD-Video players
- Elta: DVD-Video players
- Eltax: DVD-Video players
- Emerson (Funai): DVD-Video players
- Encore: DVD-Video players
- Enzer: DVD-Video players
- Esonic: DVD-Video players
- ESS Technology: DVD-Video players and WebDVD players
- Euro Asia Technologies: DVD-Video players (UK)
- Faroudja: DVD-Video players
- Finlux: DVD-Video players
- Fisher (Sanyo): DVD-Video players
- Funai (Emerson/Orion/Sylvania/Symphonic): DVD-Video players
- GE (Thomson): DVD-Video players
- Genica: DVD-Video players
- Goodmans: DVD-Video players

- GPX/Yorx: DVD-Video players
- Gradiente: DVD-Video players
- Grandin: DVD-Video players
- Great Wall: DVD-Video players (Hong Kong)
- Grundig: DVD-Video players
- Guangdong Jinzheng Digital: DVD-Video players
- Gynco: DVD-Video players
- Haier: DVD-Video players
- Harman Kardon: DVD-Video players
- Himage: DVD-Video players
- Hitachi: DVD-Video players and recorders
- Hiteker: DVD-Video players (made by VDDV)
- Homemighty: DVD-Video players
- Hoyo (Raite): DVD-Video Players
- Hyundai: DVD-Video players
- iDVDBox: Enhanced DVD-Video Players
- I-Jam: DVD-Video players
- Innovacom: PC/TV with DVD support
- Irradio: DVD-Video players
- Jasmine: DVD-Video players
- Jeutech: DVD-Video players
- JNL: DVD-Video players
- Jocel: DVD-Video players
- JVC (Victor): DVD-Video players and recorders
- Kendo: DVD-Video players
- Kennex: DVD-Video players
- Kenwood: DVD-Video players
- Keymat: DVD-Video players
- KiSS (Raite): DVD-Video players
- Kioto: DVD-Video players
- KLH: DVD-Video players
- Kones: DVD-Video players
- Konka: DVD-Video players
- Labway: DVD-Video players
- Lafayette: DVD-Video and DVD-Audio players
- Lasonic (Yung Fu): DVD-Video players
- Lawson: DVD-Video players
- Lecson: DVD-Video players

- Lector: DVD-Video players
- Legend: DVD-Video players
- Lenco: DVD-Video players
- Lenoxx: DVD-Video players
- LG Electronics (GoldStar): DVD-Video players
- Lifetec: DVD-Video players
- Limit: DVD-Video players
- Loewe: DVD-Video players
- Logix: DVD-Video players
- Lumatron: DVD-Video players
- Luxman: DVD-Video players
- Madrigal (Mark Levinson): DVD-Audio and DVD-Video players
- Magnavox (Philips): DVD-Video players
- Magnex: DVD-Video players
- Majestic: DVD-Video players
- Malata: DVD-Video players
- Manhattan: DVD-Video players
- Marantz (Philips): DVD-Audio, SACD, and DVD-Video players
- Mark: DVD-Video players
- Matsushita (Panasonic/National/Technics/Quasar): DVD-Video players and recorders, DVD-Audio players, DVD car navigation/entertainment
- Matsui: DVD-Video players
- Medion: DVD-Video players
- Memorex: DVD-Video players
- Meridian: DVD-Video players
- Metz: DVD-Video players
- MiCO: DVD-Video players
- Microboss: DVD-Video players
- Micromega: DVD-Video players
- Minato: DVD-Video players
- Mintek: DVD-Video players
- Mishine: DVD-Video players
- Mitsubishi: DVD-Video players
- Mitsui: DVD-Video players
- Monica/Monyka (Raite): DVD-Video players
- Mossimo: DVD-Video players (China)
- Mustek: DVD-Video players

- NAD: DVD-Video players
- Nakamichi: DVD-Audio and DVD-Video players
- Napa: DVD-Video players
- NEC: DVD-RAM video camera
- Neufunk: DVD-Video players
- Nintaus (Guangdong Jinzheng): DVD-Video players
- Noriko: DVD-Video players
- Odyssey: DVD-Video players
- Olidata: DVD-Video players (Italy)
- Omni: DVD-Video players
- Onkyo: DVD-Video and DVD-Audio players
- Optics-Storage: DVD-RW video recorders (supplier)
- Optim: DVD-Video players
- Orava: DVD-Video players
- Orion: DVD-Video players
- Oritron: DVD-Video players
- Palsonic (Australia): DVD-Video players
- Panasonic (Matsushita): DVD-Video players and recorders, DVD-Audio players
- Philco: DVD-Video players
- Philips (Magnavox/Marantz/Norelco): DVD-Video players and recorders
- Phoenix: DVD-Video players
- Phonotrend: DVD-Video players
- Pioneer: DVD-Video players and recorders, DVD-Audio players, DVD car navigation/entertainment
- Primare: DVD-Video players
- Proceed: DVD-Video players
- Proline: DVD-Video players
- Proscan (Thomson): DVD-Video players
- Proson: DVD-Video players
- Proton: DVD-Video players
- Quadro: DVD-Video players
- Raite: DVD-video players (Taiwan)
- Rankarena: DVD-Video players
- RCA (Thomson): DVD-video players
- RCR: DVD-Video players (China)
- REC: DVD-Video players (UK, made by VDDV, same as APEX)

- Redstar: DVD-Video players
- Revoy (Netherlands): DVD-video players
- Roadstar: DVD-Video players
- Rotel: DVD-video players
- Rowa: DVD-Video players
- Runco: DVD-video players and changers
- Saivod: DVD-Video players
- Sampo (Afreey): DVD-Video players
- Samsung: DVD-Video players
- Samwin: DVD-Video players
- Sanyo: DVD-Video players
- SAST: DVD-Video players
- Schaub Lorenz: DVD-Video players
- Schneider: DVD-Video players
- Scott: DVD-Video players
- SEG (Yamakawa/Raite): DVD-Video players
- Sharp: DVD-Video players
- Shinco: DVD-Video players (Hong Kong)
- Shinsonic: DVD-Video players
- Singer: DVD-Video players
- Skyworth: DVD-Video players
- SMC: DVD-Video players
- Sonic Blue: DVD-Video players and combo DVD-VHS players (formerly Sensory Science and Go-Video)
- Sony: DVD-Video players and changers
- Soyea: DVD-Video players
- Spatializer Audio Laboratories: 3D audio processing
- Sublime: DVD-Video players
- Sylvania (Funai): DVD-Video players
- Symphonic (Funai): DVD-Video players
- Tatung: DVD-Video players
- Teac: DVD-Video players
- Technics (Matsushita): DVD-Video and DVD-Audio players
- Teknema (Ravisent): Web-connected DVD-Video players
- Telestar: DVD-Video players
- Tevion: DVD-Video players
- Thakral: DVD-Video players (China, Hong Kong)
- Theta: DVD-Video players

- Thomson (RCA/G.E./Proscan/Ferguson/Nordmende/Telefunken/ Saba/Brandt): DVD-Video players
- Tokai (Raite): DVD-Video Players
- Toshiba: DVD-Video players and recorders, DVD-Audio players
- Tredex: DVD-Video players
- Umax: DVD-Video players
- United: DVD-Video players
- Unity Motion: DVD-Video players
- Universum: DVD-Video players
- Venturer: DVD-Video players
- Vialta (ESS): WebDVD players
- Victor (JVC): DVD-Video players
- Vieta: DVD-Video players
- Visual Disc and Digital Video: DVD-Video players (China)
- Waitec: DVD-Video players
- Walkvision: DVD-Video players
- Wharfedale: DVD-Video players
- Wintel: DVD-Video players
- XMS: DVD-Video players
- Xwave: DVD-Video players
- Yamaha: DVD-Audio and DVD-Video players
- Yamakawa (Raite): DVD-Video players
- Yami (Raite): DVD-Video players
- Yelo: DVD-Video players
- Yukai: DVD-Video players
- Zenith (becoming a subsidiary of LG): DVD-Video players

Studios, Video Publishers, and Distributors

DVD File maintains a list of studio addresses, as well as DVD producer and distributor information.

- A2O Entertainment (wholesale distributor)
- A.D. Vision (anime)
- Acorn Media
- Aftermath Media (*Tender Loving Care*, interactive movie)
- All Day Entertainment
- Alphaville Pictures (distributed by Universal)

- Amazing Fantasy
- Amblin Entertainment (distributed by Universal)
- American Gramaphone
- American Software
- Anchor Bay Entertainment
- Animeigo
- A-Pix Entertainment
- Artisan Home Entertainment (formerly LIVE Entertainment)
- Arts & Entertainment DVD
- Atomic Video (adult)
- Avalanche
- Baby Einstein (infant development)
- Baker & Taylor (distributor)
- Beyond Music (distributor)
- Black Chair Productions (independent films)
- Black Entertainment Television (BET)
- BMG (Sonopress)
- Brentwood
- Brilliant Digital Entertainment (multipath movies)
- BroadcastDVD
- Buena Vista Home Video (Disney)
- CAV Distributing (distributor)
- Castle Music Pictures (music performance)
- Castle Home Video
- Cecchi Gori
- Celebrity
- Central Park Media
- Cerebellum (educational)
- Chesky
- Classic Records
- Columbia TriStar (Sony)
- Compact Media (distributor)
- Concert @ Home (Platinum Entertainment)
- Concorde Video (*12 Monkeys*, German)
- Corinth Films (Wade Williams Collection)
- Creative Design Art
- Criterion Collection
- DaViD Entertainment

- Delos International (mostly audio)
- Delta Entertainment
- Deluxe (distributor and replicator)
- DG Distributors (distributor)
- Diamond Entertainment (distributor)
- Digital Disc Entertainment
- Digital Leisure (formerly ReadySoft) (*Dragon's Lair*, *Space Ace*)
- Digital Multimedia
- Digital Versatile Disc
- Dimension Films (Miramax)
- Direct Source
- Direct Video Distribution (distributor, UK)
- Disney (Buena Vista Home Video, Dimension Films, Hollywood Pictures, Miramax, Touchstone)
- Dream Theater
- DreamWorks SKG
- DVD International (distributor)
- D-Vision
- Eaton Entertainment
- Elite Entertainment
- EMI Records
- E Real Biz
- Essex Entertainment
- Fantoma
- Filmways (distributor, Argentina/Spain)
- FOCUSFilm Entertainment
- Fox Lorber
- Front Row
- Full Moon Pictures
- Gainax (anime)
- General Media Communications (Penthouse) (adult)
- Goldhil Home Media
- Goodtimes Entertainment
- Gramercy Pictures (distributed by Universal)
- Hallmark Home Entertainment (Artisan)
- HBO Home Video (Warner)
- HODIE (multimedia recording label)
- Hollywood Pictures (Disney, folded into Touchstone)

- Hot Body International (adult)
- Ice Storm Entertainment (distributor, Germany)
- Ideal Entertainment
- Image Entertainment (distributor)
- Impressive (adult)
- IndieDVD (publisher; alliance of independent filmmakers)
- Ingram (distributor)
- Key East
- King's Road (distributed by Trimark)
- Kino International
- Laserdisc Entertainment (adult)
- Laserlight
- Lee & Lee Films
- Leo Films
- Living Arts (health)
- LucasFilm (distributed by Twentieth Century Fox or Paramount)
- Lucerne Media (educational)
- Lumivision (distributed by SlingShot)
- Lyric
- MacDaddy
- Madacy
- Magic Lantern
- Marin Digital (Your Yoga Practice)
- Master Tone
- MCA (Universal)
- MCA Music
- Media Galleries
- Media Group (distributor)
- Metro Global Media (adult)
- Metromedia
- MGM/UA (Warner)
- Mill Reef (*Earthlight*)
- Miramax Films (Disney)
- Monarch Home Video
- Monterey
- MPI Home Video
- MTI
- Multimedia 2000 (aka M-2K)

- Music Video Distributors (distributor)
- N2K Music
- Navarre (distributor)
- NET TEN (distributor)
- Nettwerk Productions
- New Horizons Home Video
- New Line (Warner)
- New Video Group
- New Vision
- New York Entertainment
- NuTech Digital (adult)
- October Films (Universal)
- Opera World
- Orion Pictures (MGM, some older DVD titles distributed by Image and Criterion)
- Overseas Filmgroup (distributor, partner with Image)
- Pacific Digital
- Palm Pictures
- Panasonic Interactive Media (defunct)
- Panorama
- Paramount Home Video (owned by Viacom)
- Parasol
- Passport Video
- Phantom Video
- Picture This Home Video
- Pioneer Entertainment (distributor)
- Platinum
- Playboy Home Video
- PM Entertainment
- Polygram (Philips partner)
- Pony Canyon (Japan)
- PPI Entertainment
- Private Media Group (adult)
- Pro7 Home Entertainment (Germany)
- Program Power
- Real Entertainment
- Red Distribution (distributor)
- Renegade

- Republic Pictures (defunct, distributed by Artisan)
- Rhino Home Video
- Roadshow Entertainment (Australia)
- Roan Group
- Rykodisc
- Samsung Entertainment Group
- Shanachie
- Showtime
- Simitar Entertainment
- Sierra Vista Entertainment (Innovacom)
- Silver Screen
- SlingShot (acquired Lumivision titles)
- Sony Music Entertainment
- Sony Pictures (Columbia, Epic, Sony Music, Sony Wonder, TriStar)
- Sony Wonder (kids)
- Steeplechase
- Sterling Home Entertainment
- Super Digital Media
- SyCoNet.com (distributor, anime)
- Synapse Films
- Tai Seng
- Technicolor (distributor and replicator)
- Tempe Entertainment
- Thakral (distributor; Hong Kong, China)
- Toho (Japan)
- Tone Home Video
- Toshiba EMI
- Touchstone (Disney)
- Trimark Pictures
- Troma Entertainment
- Turner Home Entertainment
- Twentieth Century Fox Home Entertainment
- Unapix Entertainment
- United American
- United Artists (MGM)
- Universal Studios Home Video (owned by Seagram)
- USA
- U.S. Laser

- Valley Media (distributor)
- VCA Interactive (VCA Pictures, VCA Labs; adult)
- VCI Home Video
- Ventura
- Victor Entertainment (JVC)
- Victory
- Video Watchdog
- Video One Canada (distributor)
- Vidmark
- Vista Street
- Vivid Video (adult)
- Walt Disney Pictures
- Warner Bros. Records/Warner Music (Toshiba partner)
- Warner Home Video (Toshiba partner)
- Waterbearer Films
- WIT Entertainment (distributor)
- WGBH
- WWF Home Video
- Wolfe
- World Video
- Xenon
- Xoom
- York

Hardware and Computer Components

- Acer Laboratories: DVD decoder/controller chips
- Advent: DVD-ROM-equipped computers
- Alliance Semiconductor: display adapters with hardware acceleration for DVD playback
- Allion: DVD mirroring servers
- AMLogic: DVD player chipset
- Analog Devices: 192-kHz/24-bit audio DAC
- Apple: DVD-ROM- and DVD-RAM-equipped computers, playback hardware and software (QuickTime)
- ASACA: DVD-RAM towers
- AST: DVD-ROM-equipped computers (with MMX-based playback software)
- ASM: DVD jukeboxes

- ATI Technologies: display adapters with hardware acceleration for DVD playback
- Avid Electronics: DVD decoder/controller chips
- Axis Communications: DVD-ROM storage servers
- Bridge Technology: optical pickup assemblies
- Canopus: DVD-RAM video archiving.
- CD Associates: Software and hardware for production and testing.
- CEI: DVD playback hardware and software
- Cirrus Logic: MPEG-2 encoder/decoder chips
- CMC Magnetics: recordable discs
- Compaq: DVD-ROM-equipped computers
- Creative Technology: DVD-ROM and DVD-RAM upgrade kits, DVD decoder software
- Cygnet: DVD-RAM jukeboxes
- DIC (Dainippon Ink and Chemicals): ink, organic pigments, thermosetting resin
- Dave Jones Design: controllers for industrial DVD players
- Diamond Multimedia: DVD upgrade kit (Toshiba drive)
- Digimarc: watermarking technology
- Digital: DVD software playback (for Alpha workstations), DVD encoder chips
- Digital Stream: optical pickup assemblie
- Digital Video Systems: DVD-ROM drives
- Disc, Inc.: DVD-RAM jukeboxes.
- DSM: DVD jukeboxes
- DVDO: video deinterlacing chips
- DynaTek: DVD upgrade kit
- EPO Technology: DVD-ROM drives
- Escient: DVD-ROM changer
- ESS Technology: playback chipset, player reference design
- Fantom Drives: DVD-RAM and DVD-ROM kits
- Fujitsu: DVD-ROM-equipped computers
- Gateway: DVD-ROM-equipped computers
- Genesis Microchip: video chips (progressive-scan, scaling)
- Granite Microsystems: IEEE-1394 DVD-ROM drives
- Harman Int.: DVD jukebox
- Hitachi: DVD-ROM drives, DVD-RAM drives, decoder chips
- Hi-Val: DVD playback hardware (upgrade kit)

- Hyundai: DVD decoder chips
- IBM: DVD-ROM-equipped computers, decoder chips
- I-Jam: DVD-ROM drives
- Imation: DVD-RAM media.
- Inaka: DVD jukebox software
- Infineon: DVD reader circuitry
- Innovacom: DVD encoder and decoder systems
- Intel: DVD playback hardware (MMX) and software
- Interactive Seating: Battle Chair
- I/OMagic: IEEE-1394 DVD-ROM drives
- JVC: DVD-ROM drives, DVD-RAM jukebox
- Kasan: decoder hardware
- KOM: DVD-RAM changer
- LaCie: DVD-RAM drives
- Leitch: DVD-RAM video recording
- LG Electronics: DVD-ROM drives
- LSI: DVD encoder and decoder chips (acquired C-Cube)
- Luminex: Unix software for DVD-based archiving and duplication
- LuxSonor: DVD playback chips
- Margi: DVD decoder card for notebook PCs
- Matrox: display adapters with hardware acceleration for DVD playback
- Matsushita (Panasonic): DVD-ROM drives, DVD-RAM drives, upgrade kits, DVD/Web integration, DVD-RAM still-image recorder
- Media100: DVD authoring tools, DVD playback hardware and software
- Mediamatics: DVD playback software and hardware
- Medianix: Dolby Digital decoder hardware with Spatializer 3D audio
- Memorex: DVD-ROM drives
- Microboards: DVD drive (VAR)
- Microsoft: DVD playback support (DirectShow) and player applications
- Microtest: DVD-ROM jukeboxes
- Mitsubishi: DVD players, DVD-ROM drives
- Motorola: DVD decoder chips
- National Semiconductor: DVD playback and reference designs
- Number 9: display adapters with hardware acceleration for DVD playback

- Nuon Semiconductor: DVD playback reference platform (Nuon)
- NEC: DVD-ROM drives
- Net TV: DVD-ROM PC for home entertainment
- NSM: DVD-ROM jukebox, DVD-RAM jukebox
- Oak Technology: DVD playback hardware and software
- OTG Software: DVD jukebox software
- Packard Bell: DVD-ROM-equipped computers
- Philips: DVD-ROM drives, DVD+RW drives, decoder chips
- Pioneer: DVD-ROM drives, DVD-R drives, DVD-RW video recorders
- Plasmon Data: DVD-RAM jukebox
- Procom: DVD-ROM jukebox
- Ricoh: DVD-ROM/CD-RW drives
- Rimage: DVD duplication and printing equipment
- RITEK: DVD-R, DVD-RAM
- S3: display adapters with hardware acceleration for DVD playback
- Samsung: DVD-ROM drives and DVD-ROM-equipped computers
- Spectradisc: limit-play technology
- STMicroelectronics (formerly SGS-Thomson): DVD decoder chips
- SICAN: DVD decoder chips
- Sigma Designs: DVD playback hardware
- Software Architects: DVD-recordable utilities for UDF and Mt. Rainier writing
- Sonic Solutions: DVD-Video decoding software (acquired a portion of Ravisent, formerly Quadrant International)
- Sony: DVD-ROM drives, DVD-ROM-equipped computers
- ST Microelectronics: DVD decoder chips (acquired a portion of Ravisent, formerly Quadrant International)
- STB Systems: DVD playback hardware (upgrade kit)
- Technovision: Controllers and synchronizers for consumer and industrial DVD players
- TDK: blank DVD-RAM discs
- Toshiba: DVD-ROM drives, DVD-ROM-equipped computers, DVD-RAM drives
- Tracer Technologies: DVD jukebox software and DVD recording software (Unix)
- TribeWorks: custom player software
- Trident Microsystems: DVD decoder chips, DVD-accelerated video controller chips
- Truevision: DVD playback software (Microsoft Active Movie 2.0)

- Verbatim Australia (ActiveMedia): DVD playback hardware (upgrade kit)
- VisionTech: MPEG-2 encoder/mulitplexer
- Wired: DVD playback hardware and software (acquired by Media 100)
- X-10.com: (wireless DVD transmitter)
- Xing: DVD playback software
- Yamaha: AC-3 decoder chips
- Zen: multi-beam DVD reading technology
- Zoran/CompCore: DVD software and hardware playback, DVD decoder chips

Computer Software Titles on DVD-ROM

- 2 Way Media: Launch
- Access Software: Overseer, Tex Murphy
- Acclaim Entertainment: Reah
- Accolade: Jack Nicklaus 4, Family Spectacular
- Action Zone: games
- Activision (Quicksilver): Muppet Treasure Island, Spycraft: The Great Game, Zork: The Grand Inquisitor
- Aftermath Media: Tender Loving Care
- ALLDATA: automotive information databases
- Aludra: Beat 2000 DVD, Language Tutor DVD, Virtual Makeover DVD
- Apple Computer: Mac OS Anthology (available to developers only)
- BBC Interactive
- Black Isle Studios (Interplay): Baldur's Gate
- Broderbund: Riven, PrintMaster Platinum, ClickArt 300,000.
- Byron Preiss/Simon & Schuster: The Timetables of Technology
- ComChoice: Marketing, sales, and training
- Creative Multimedia: Billboard Music Guide, Blockbuster Entertainment Guide to Movies and Video
- Creative Wonders (The Learning Company): Schoolhouse Rock, Sesame Street, Wide World of Animals
- DeLorme: AAA Map'n'Go DVD Deluxe
- Data Becker: Clipart Collection, Sound Collection
- Digital Directory Assistance: PhoneDisc PowerFinder USA One
- Digital Versatile Disc: Shaodan
- Digital Leisure: Dragon's Lair, Hologram Time Traveler, Space Ace

- Discovery Channel: Leopard Son/Animal Planet, Connections
- Dorling Kindersley
- Electronic Arts: Wing Commander IV
- Electronic Publishing Association: LANGMaster Collins COBUILD Student Dictionary
- EuroTalk Interactive: Language Learning
- Firebrand: Lost in Crazy Town
- genX Software: Dead Moon Junction
- Global Star Software: 100 Great Action Arcade Games, Excessive Speed, Gubble, 303 Professional Legal Forms
- Graphix Zone
- Grolier: Multimedia Encyclopedia
- GT Entertainment: Forrest Gump, Reah
- Hachette Multimedia: Hachette Encyclopedia
- IBM Interactive Media: The Pistol: The Birth of a Legend
- Index+: Dracula Resurrection, Dracula the Last Sanctuary, Louvre the Final Curse
- Interactual Technologies: Star Trek VideoSaver
- Interplay: Baldur's Gate, Starfleet Academy
- Into Networks: PlayNow (unlockable games)
- IVS: The Union Catalogue of Belgian Research Libraries
- Japan Travel Bureau: DVD-Web product
- Kunskapsforlaget (Sweden): Focus Encylopedia
- The Learning Company (SoftKey): Battles of the World, Clickart, Digital Library, The Genius of Edison, National Geographic, Printmaster 7.
- Liris (Havas) Interactive: Découvertes (Junior Discovery)
- Magnum Design
- Mechadeus: The Daedalus Encounter
- MediaGalleries: Multimedia Bach
- MediaOne: VersaDisc
- Microsoft: Encarta, MSDN/TechNet, Works Suite
- Mill Reef: Earthlight, Coral Sea Dreaming
- Mindscape
- Mitchell Repair Information Company: ON-DEMAND
- Monolith: Claw
- Montparnasse Multimedia: Microcomsos, Voyage to the land of the Pharaohs

- Multimedia 2000 (aka M-2K, formerly Multicom): Birds of the World; Bubblegum Crisis; HomeDepot's Home Improvement 1-2-3; Warren Miller's Ski World '97; Exploring National Parks; Great Chefs, Great Cities; Better Homes and Gardens Cool Crafts
- Natif
- NB Digital/Mill Reef: Earthlight
- Not A Number: Blender
- Oeil Pour Oeil: Death Dealer
- Organa: The Book of Lulu
- Pro CD: Select Phone
- Project Two Interactive: Reah (distributed by GT in U.S., Acclaim in UK and Ireland)
- Psygnosis: Lande
- Red Orb Entertainment:
- Sega: 4 game/instruction titles to be released in early 1997
- Sierra Online
- Sumeria: Vanishing Wonders of the Sea, Wild Africa
- SuperZero: adult DVD-Video
- SuSE: SuSE Linux 6.3
- TerraGlyph Interactive Studios: Buster and the Beanstalk (Tiny Toons)
- Torus Games
- Tsunami: Crazy 8's, Silent Steel, Silent Steel II
- VR Sports (Interplay): Virtual Pool
- Warner Advanced Media
- Westwood Studios: Command & Conquer
- Xiphias: Encyclopedia Electronica
- Zombie VR Studios: Liberty

Where Can I Buy (or Rent) DVDs and Players?

(See "How Much Do Discs Cost?" in Chapter 1, "General DVD," for price comparisons and coupons.)

- 800.com (players)
- 999Central (DVDs for shipping and handling cost only)
- A&B Sound (Canada)
- abcDVD (UK, region 1)
- abt Electronics (players)

- AccessDVD.com
- Ace VCD DVD (Hong Kong/anime)
- Airplay (Japan, region 2)
- All DVD Movies (DVDs)
- AllCheapMusic.com (DVDs for $10 or less)
- Amazon.com (players and DVDs)
- Amazon.co.uk (UK; players and DVDs)
- AnimeNation (DVDs)
- Anime Depot (DVDs)
- Asian Xpress (Hong Kong films)
- Bargainflix (DVDs)
- Best Buy (players and DVDs)
- Best Buy Movie (Germany; DVDs)
- Bensons World (UK; players)
- Beyond Music (DVDs
- Big Emma (used DVDs)
- BigStar (players and DVDs)
- BigWheelOnline.com (DVDs; $1 shipping worldwide)
- BlackStar (UK, region 2 DVDs; free shipping worldwide)
- Blockbuster (rental and sales of DVDs)
- Brainplay.com (DVDs)
- Buy.com (players and DVDs)
- C&L Internet Club (Canada; DVDs)
- CD JAPAN (Japan, region 2)
- CDNOW (DVDs)
- CDRealm (Switzerland)
- Columbia House (DVD mailorder club)
- Consumer Direct Warehouse (players)
- Critics' Choice Video (DVDs)
- DeepDiscountDVD.com
- DeVoteD (Australia, region 4 DVDs)
- Digibuster Media (online rental)
- Digital Entertainment (Indian films)
- Digital Eyes (DVDs)
- Digital Playtime (Australia, region 4)
- Digitallageret.com (Asian imports)
- The Digital Shop (Greece)
- Direct Video

- Disc and Picture Company (Australia)
- discShop.com (UK, region 1 and 2)
- DVDCity
- DVD City (Australia)
- DV Depot
- DVD Domain
- DVD Empire
- DVDIt Italia (Italy)
- DVD North (Canada)
- DVDONE
- DVD Overnight (online rental)
- DVD Palace (formerly Liquidata)
- DVD Planet (formerly Ken Crane's, now a division of Image Entertaiment)
- DVDPlus (Europe)
- DVD Rent (Australia, sales and online rental)
- DVDshoppingCenter (region 2)
- The DVD Movie Store (Australia, offline rentals)
- DVDstreet (region 2)
- DVD Supercenter.com (adult)
- DVD titlewaves (discs and players)
- DVD VideoPlanet (New Zealand, regions 1 and 4)
- DVD Wave
- DVD World (UK, region 2)
- DVD World (New Zealand, regions 1 and 4)
- DVD Zone 2 (region 2)
- eBay (buy and sell new and used DVDs)
- Elvic (Netherlands)
- EntertainmentStudios.com (DVDs)
- Evolution Audio & Video
- Express.com
- Fantastic Movies (Switzerland)
- FeatureDVD
- Fotosound (UK)
- Gamestech.com (multi-region players)
- German Music Express (Germany)
- GreenCine (online rental of rare and alternative titles)
- Just Watch It (regions 1 and 2)

- Karaoke - Show (Switzerland)
- LADA Universal (regions 1 and 2, new and used)
- Laser Corner (Greece)
- Laserdisc DVD Outlet
- Laser Discovery (online rental, Hong Kong movies)
- The LaserDisc Division
- Laserdisc House (UK, regions 1 and 2)
- Laser's Edge
- Laservisions Direct
- LearningStore.co.uk (educational and non-violent DVDs)
- GoDVD (UK, regions 1 and 2)
- Half.com (used discs and players)
- Hastings Entertainment (buy or rent DVDs)
- Hifi.com (players)
- HKFlix.com (Asian DVDs)
- InsideDVD (free disc subscription)
- Hollywood Video (rental)
- Ken Crane's
- Kotiteatteri (Finland)
- Media Play
- MegaDVD
- MovieClubOnline (discount video rentals)
- MovieGallery.com (new and used movies and games)
- Musicland
- NetFlix (online rental, monthly fee)
- North American DVD (retail and wholesale)
- On Cue
- OneCall (players)
- OZDVD Warehouse (region 4)
- Reel.com (no longer sells discs)
- Reg2.net (Spain)
- Rent A DVD (online rental, Switzerland)
- Ro-Disc (Netherlands, regions 1 and 2)
- RPM Records (rare discs)
- Sam Goody
- Second Chance DVD (used)
- Shopping.com
- Shopping Matrix (South Africa, region 2)

- Sony Music Direct
- Stardust DVD (Puerto Rico)
- Starship Industries
- SublimeDigital.com (players and drives)
- SVS (UK, region 2)
- Swinging Planet (UK, cult video; region 2)
- TLA Video
- Trans World Entertainment (TWEC)
- Universe of Entertainment (Switzerland)
- VideoCave
- VideoLtd.com
- Virgin Megastore
- Xchangecity (trade DVDs with other members)
- (Disclosure: Some of the links above include affiliate program information that may result in a commission to Jim.)

Where Can I Buy Blank Recordable DVDs?

Important note: With blank DVDs the adage "you get what you pay for" is usually true. Cheaper discs are more likely to produce errors when burning and are less compatible with players.

- 800 CDR
- Buy-DVD-R.com
- CD-DVD-Supplies.com
- CD-Recordable.com
- J&R Electronics
- Meritline.com
- Memorex
- Pro Tape Northwest
- Shop4tech.com

Where Can I Get More Information About DVD?

A Few of the Top DVD Info Sites

- Robert's DVD Info (tons of links to news articles and other pages, but not updated for over a year)
- DVDPhD www.dvdphd.com (DVD tech support)

- The Digital Bits www.thedigitalbits.com (top DVD news site)
- DVDFile www.dvdfile.com (another good DVD news site)
- DVD Review www.dvdreview.com (DVD news and production information)
- DVDAnswers www.dvdanswers.com (general DVD info site)
- Home Theater Forum www.hometheaterforum.com (general DVD discussions)
- TheDVDPlayer www.thedvdplayer.com (immense collection of links to other DVD pages)
- Chad Fogg's DVD technical notes www.mpeg.org/~tristan/MPEG/DVD/ (from 1996)
- Quantel Digital Fact Book (digital video info and glossary) www.quantel.com/dfb
- DVD for not-so-Dummies, from Technicolor www.technicolor.com/services/DVD2000v1.pdf
- DVD White Papers, from Sonic Solutions www.sonic.com/tech_whitepapers.html
- Disctronics' (Graham Sharpless's) DVD Technology pages www.discusa.com/dvd/dvdmain.htm
- Tristan's MPEG Pointers and Resources www.mpeg.org
- DVD discussion list. Send "subscribe DVD-L <your name>" to listserv@listserv.temple.edu
- For details on YUV, RGB, YCbCr, etc., read Charles Poynton's Color FAQ (or buy his book).

DVD Utilities and Region-Free Information

(See "What Are 'Regional Codes,' 'Country Codes,' or 'Zone Locks'?" in Chapter 1 for more information about regions.)

- DVD Infomatrix www.inmatrix.com (a wealth of information about DVD PCs)
- MPEGX www.mpegx.com (PC utilities for video and audio, more)
- DVDSoft.net www.dvdsoft.net (PC utilities, more)
- Doom9 www.doom9.net (PC utilities for DVD backup)
- Visual Domain www.visualdomain.net (PC utilities, including *Drive Info*)
- DVDCity www.dvdcity.com/codefree/codefree-dvd-info.html (code-free DVD player FAQ)
- Code Free DVD www.codefreedvd.com (region-free DVD players)

- Region Free DVD www.regionfreedvd.net (region workarounds for players and PCs)
- RegionFreeDVDPlayes regionfreedvdplayers.com
- ZoneFreeDVD zonefreedvd.com
- dvdkits.com www.dvdkits.com (modification chips for DVD players)
- DVD Upgrades www.dvdupgrades.ch (region-free DVD players and modification chips)
- DVDoverseas www.dvdoverseas.com (region-free DVD players)
- Link Electronics www.linkonline.co.uk (region-free DVD players and upgrades)
- Techtronics www.techtronics.com (region-free DVD players and upgrades)
- Upgrade Heaven www.homecinemaheaven.com (region-free DVD players)
- Eric's DVD Information www.brouhaha.com/~eric/video/dvd (tech info on early players)
- Google Deja Usenet Archive www.deja.com (search the rec.video.dvd and alt.video.dvd newsgroups)
- The Mac DVD Resource www.wormintheapple.gr/macdvd/?action=intro (region-free info for Macs)
- RipDifferent Forum www.ripdifferent.com (discussion of audio and video ripping on Macs)
- PowerBook DVD Source http://www.dfbills.com/powerbook/dvd.html (info about DVD on Macs)

Information and Discussion Groups for DVD Authors

- DVD Made Easy dvdmadeeasy.com (tutorials, forums, and other resources; fee-based)
- EZ DVD Advisor www.ezdvdadvisor.com (forums and other resources)
- DVD Developer Club at Yahoo clubs.yahoo.com/clubs/dvddeveloper (discussion of authoring techniques and problems)

DVD Info for Specific Regions

- uk.media.dvd FAQ www.dvd.reviewer.co.uk/umdvdfaq/
- UK DVD FAQ movieuk.com/dvdfaq.htm (not updated since 12/98)

- DVD Debate www.dvddebate.com (news, info, and user discussions; mainly region 2)
- DVD Times www.dvdtimes.co.uk (news, info, and reviews; mainly region 2)
- DVD Reviewer www.dvd.reviewer.co.uk (news, info, and reviews; mainly UK)
- DVDLink www.dvdlink.co.uk (news and links to hundreds of other DVD sites)

DVD Info in Other Languages

- DVDUpdate www.dvdupdate.nl (Dutch)
- dvdfr.com www.dvdfr.com (French)
- Area DVD www.areadvd.de (German)
- DVD-Inside www.dvdinside.ch (German)
- DVDPrime www.dvdprime.com (Korean)
- dvdnett.no www.dvdnett.no (Norwegian)
- DVD'mension dvd.wp.pl (Polish)
- DVDSpecial www.dvdspecial.ru (Russian)
- Audio Video Cine en Casa club.idecnet.com/~modegar/ (Spanish)

Books About DVD

- *DVD Demystified*, by Jim Taylor
- *DVD Authoring and Production*, by Ralph LaBarge
- *Desktop DVD Production*, by Douglas Dixon
- *From VHS to DVD*, by Mark-Steffen Goewecke
- *CD-R/DVD Disc Recording Demystified*, by Lee Purcell
- *DVD Production*, by Phil De Lancie and Mark Ely

What's New with DVD Technology?

June 2003

There are rumors that there's a sixth HD format in the works based on the +RW format.

March 2003

There are now at least 5 candidates for high-definition DVD. (See "What About the HD-DVD and Blue Laser Formats?" in Chapter 3, "DVD Technical Details," for details).

- HD-DVD-9 (aka HD-9).
- Advanced Optical Disc (AOD).
- Blu-ray (BD).
- Advanced Optical Storage Research Alliance (AOSRA), Blue-HD-DVD-1.
- AOSRA Blue-DVD-DVD-2.

June 2002

Philips demonstrated a blue-laser miniature pre-recorded optical disc. The 3-cm (1.2-inch) disc holds 1 Gbyte of data. The prototype drive to read the disc measured 5.6 x 3.4 x 0.75 cm (2.2 x 1.3 x 0.3 inches).

February-March 2002

A group of 9 companies announced February 19th a new high-density recordable DVD standard, known as Blu-ray. At the DVD Forum general meeting in March, the Forum announced that it will investigate next-generation standards to choose the best one. Since the 9 companies are all members of the DVD Forum, it's likely that Blu-ray will eventually be approved by the Forum.

Also at the March meeting, the Forum announced that according to AOL Time Warner's request it will work on a standard for putting high-definition video on existing DVDs. The format is being called "HD-DVD-9." (See "What About the HD-DVD and Blue Laser Formats?" in Chapter 3 for details).

Chapter 7

Leftovers

Notation and Units

There's an unfortunate confusion of units of measurement in the DVD world. For example, a single-layer DVD holds 4.7 billion bytes (G bytes), not 4.7 gigabytes (GB). It only holds 4.37 gigabytes. Likewise, a double-sided, dual-layer DVD holds only 15.90 gigabytes, which is 17 billion bytes.

The problem is that the SI prefixes kilo, mega, and giga normally represent multiples of 1000 (10^3, 10^6, and 10^9), but when used in the computer world to measure bytes they generally represent multiples of 1024 (2^{10}, 2^{20}, and 2^{30}). Both Windows and Mac OS list volume capacities in "true" megabytes and gigabytes, not millions and billions of bytes.

Most DVD figures are based on multiples of 1000, in spite of using notation such as GB and KB that traditionally have been based on 1024. The "G bytes" notation does seem to consistently refer to 10^9. The closest I have been able to get to an unambiguous notation is to use kilobytes for 1024 bytes, megabytes for 1,048,576 bytes, gigabytes for 1,073,741,824 bytes, and BB for 1,000,000,000 bytes.

This may seem like a meaningless distinction, but it's not trivial to someone who prepares 4.7 gigabytes of data (according to the OS) and then wastes a DVD-R or two learning that the disc really holds only 4.3 gigabytes! (See "What Are the Sizes and Capacities of DVD?" in Chapter 3, "DVD Technical Details," for a table of capacities.)

Here's an analogy that might help. A standard mile is 5,280 feet, whereas a nautical mile is roughly 6,076 feet. If you measure the distance between two cities you will get a smaller number in nautical miles, since nautical miles are longer. For example, the distance from Seattle to San Francisco is about 4,213,968 feet, which is 798 standard miles but only 693 nautical miles. DVD capacities have similarly confusing units of measurement: a billion bytes (1,000,000,000 bytes) or a gigabyte (1,073,741,824 bytes). DVD capacities are usually given in billions of bytes, such as 4.7 billion bytes for a recordable disc. Computer files are measured in gigabytes. Unfortunately, both types of measurements are often labeled as "GB." So a 4.5GB file

(4.5 gigabytes) from a computer will not fit on a 4.7-GB disc (4.7 billion bytes), since the file contains 4.8 billion bytes.

To make things worse, data transfer rates when measured in bits per second are almost always multiples of 1000, but when measured in bytes per second are sometimes multiples of 1024. For example, a 1x DVD drive transfers data at 11.08 million bits per second (Mbps), which is 1.385 million bytes per second, but only 1.321 megabytes per second. The 150 KB/s 1x data rate commonly listed for CD-ROM drives is "true" kilobytes per second, since the data rate is actually 153.6 thousand bytes per second. This book uses kbps for thousands of bits/sec and Mbps for millions of bits/sec (note the small "k" and big "M").

In December 1998, the IEC produced new prefixes for binary multiples: kibibytes (KiB), mebibytes (MiB), gibibytes (GiB), tebibytes (TiB), and so on. These prefixes may never catch on, or they may cause even more confusion, but they are a valiant effort to solve the problem. The big strike against them is that they sound rather silly.

Acknowledgments

This book and the online FAQ has been written and is maintained by Jim Taylor. The following people contributed to early versions of the DVD FAQ. Their contributions are deeply appreciated. Information has also been taken from material distributed at the April 1996 DVD Forum, May 1997 DVD-R/DVD-RAM Conference, and October 1998 DVD Forum Conference.

Robert Lundemo Aas

Adam Barratt

David Boulet

Espen Braathen

Wayne Bundrick

Irek Defee

Roger Dressler

Chad Fogg

Dwayne Fujima

Robert "Obi" George

Henrik "Leopold" Herranen

Kilroy Hughes

Mark Johnson

Ralph LaBarge

Martin Leese
Dana Parker
Eric Smith
Steve Tannehill
Geoffrey Tully

Thanks to Videodiscovery for hosting the online FAQ for the first two and a half years.

Index

About the Author

Jim Taylor is chief of DVD technology and general manager of the Advanced Technology Group at Sonic Solutions, the leading developer of DVD authoring systems. The author of McGraw-Hill's best-selling *DVD Demystified*, he was called "a minor tech legend" by E! Online. Creator of the legendary DVD FAQ web site at dvddemystified.com, he also writes articles and columns about DVD, serves as president of the DVD Association, and sits on several advisory boards of leading-edge companies in the DVD industry. Jim was named one of the 21 most influential DVD executives by *DVD Report*, and was an inaugural inductee into the Digital Media Hall of Fame. He lives in Seattle, Washington.